AF754806

QUANTUM DOT PHOTOVOLTAICS

QUANTUM DOT PHOTOVOLTAICS

Edited by

Nelson Bolívar

ARCLER PRESS

www.arclerpress.com

Quantum Dot Photovoltaics

Nelson Bolívar

Arcler Press
2010 Winston Park Drive,
2nd Floor
Oakville, ON L6H 5R7
Canada
www.arclerpress.com
Tel: 001-289-291-7705
001-905-616-2116
Fax: 001-289-291-7601
Email: orders@arclereducation.com

© 2019 Arcler Press

ISBN: 978-1-77361-093-1 (Hardcover)

This book contains information obtained from highly regarded resources. Reprinted material sources are indicated and copyright remains with the original owners. Copyright for images and other graphics remains with the original owners as indicated. A Wide variety of references are listed. Reasonable efforts have been made to publish reliable data. Authors or Editors or Publishers are not responsible for the accuracy of the information in the published chapters or consequences of their use. The publisher assumes no responsibility for any damage or grievance to the persons or property arising out of the use of any materials, instructions, methods or thoughts in the book. The authors or editors and the publisher have attempted to trace the copyright holders of all material reproduced in this publication and apologize to copyright holders if permission has not been obtained. If any copyright holder has not been acknowledged, please write to us so we may rectify.

Notice: Registered trademark of products or corporate names are used only for explanation and identification without intent of infringement.

Arcler Press publishes wide variety of books and eBooks. For more information about Arcler Press and its products, visit our website at www.arclerpress.com

ABOUT THE EDITOR

Nelson Bolivar has a PhD. in physics from the University of Lorraine in France finished in 2014. His expertise is in quantum systems and condensed matter. His interest includes spintronic devices and correspondences between general relativity and condensed matter. He is currently an associate professor at the Central University of Venezuela.

TABLE OF CONTENTS

LIST OF FIGURES

As produced films having oleic acid capping group reveal an average QD-QD space of about 1.8 nm while ethylenediamine treatment can result in the reduced distance, i.e., up to < 0.4 nm. The immediate advantage of a reduced QD-QD distance is better QD-QD electronic interaction. This phenomenon is graphically illustrated in the lower panel which shows the overlapping electronic wave-functions for excited electrons between adjoining QDs.

Figure 5.2. The architecture of device and EQE values >100% for solar cells containing PbSe quantum dots. Energy landscape of a solar cell developed from PbSe QDs that have been treated by 1,2-ethanedithiol (EDT) or hydrazine (HDZ). The variation in doping density in QD film and an electron acceptor (ZnO) creates a depletion region inside the QD layer, which impels charge extraction after the generation of free carriers. (B) External quantum efficiency (EQE) of the same PV device architecture exhibiting the high-energy region for 3 different sizes of PbSe QD. All sizes of QD show the generation of more than single carrier produced for each incident photon at $hv_{photon} = 3.3$ eV.

Figure 5.3. Internal quantum efficiencies (IQEs) of devices created from PbSe nanorods of 3 different band gaps (indicated by solid colored lines) and peak quantum yields of devices comprising of spherical PbSe QD (indicated by squares). The MEG quantum yields of nano-rods (based on PbSe) in solution and of sphere-shaped PbSe QDs in the solid state are indicated as crosses and circles, respectively.

Figure 5.4. Proof of carrier multiplication in solar cells containing PbTe quantum dot. (A) External and (B) internal quantum efficiencies of solar cells based on PbTe QD. The architecture of the device is sketched in the inset of (A). IQE model (hv) and IQE Exp (hv) are the internal quantum efficiencies determined by transfer matrix modeling and by reflectance measurements, respectively. (C) Conduction of pump-probe transient absorption experiment on films of PbTe QDs signifying a transiently stable carrier population at around 2.8 eV. (D) The response of photocurrent as a function of time delay in between excitation pulse and succeeding low-energy push pulse, which re-energizes the new carrier population.

Figure 5.5. Potential methods for upgrading state-of-the-art solar cells utilizing carrier multiplication in quantum dots. Schematics of (A) a MEG-enhanced, interdigitated bottom-contact solar cell. High-energy photons (blue arrows) are absorbed in the thin QD layer processed on top of a bulk semiconductor where they can undergo MEG. The carriers from the generated multi-exciton state are subsequently transferred to the low bandgap semiconductor and can contribute to the photocurrent. Meanwhile, photons of lower energy are more likely to pass through the QD film due to the lower absorption cross-section and can be absorbed and converted into charge carriers in the bulk semiconductor. (B) A device concept based upon a nanostructured hetero-interface between QD absorber and metal oxide acceptor. Here most multiple-exciton states are generated close to an ionizing interface. Both an efficient harvest of multi-exciton states and long optical path lengths for absorbing solar photons are possible.

LEDs also result in the generation of an electric field across the QDs which can affect the radiative and non-radiative recombination rates of excitons produced in the QD-LED.

Figure 7.2. Schematic illustration of the impacts of the QD structure on the Auger (ka) or trap-assisted (kt) non-radiative recombination rates, i.e., negative impact (red-cross) or positive impact (green check). Organization of the QD structures is based on the extent of their electronic confinement and surface passivation concerning a CdSe core (present in the upper left corner of the image). All quantum dot structures are demonstrated along with their energy band diagrams exhibiting the wave functions of holes (red) and electrons (blue).

Figure 7.3. (a) Experimental measurements for the photo-luminescent QY (quantum yield) and (b) calculations for tight binding of the elements in an optical matrix concerning the electric field (F) experienced by the QDs. (c) Schematic demonstrations for wave-functions of holes and electrons in CdS/CdSe and CdS/ZnSe QDs with and without the stimulus of an electric field which shows that relaxed confinement is undesirable in the existence of an electric field, as it causes a decline in the radiative rates because of the exciton polarization.

Figure 7.4. Schematic illustration of the challenge encountered in the selection of a quantum dot emitter for LED (light-emitting diode) applications. QDs with lower electronic confinement possess low non-radiative Auger recombination, but also show decreased radiative recombination in the existence of an electric field. On the other hand, a quantum dot with high electronic confinement exhibits high non-radiative Auger recombination and lower luminescence quenching induced by an electric field.

Figure 7.5. 2 possible approaches to circumvent luminescence quenching caused due to the electric field. (a) A dopant has the potential to provide a luminescent state which can be less sensitive to higher electric fields. Mn+ is 1 common type of dopant in QDs which emits the red light (solid line) even if the parent QD has a high bandgap, as shown in the absorption onset (dashed line) at 450 nm. (b) Embedding quantum dots in a host (parent) material with high dielectric constant (ε) can significantly diminish the effective electric field experienced by the excitons (Fex). For instance, an emissive layer containing a TiO2 matrix with a 20% quantum dot volume fraction decreases the effective electric field experienced by the quantum dots by a factor of 10 (or slightly more), as compared to a pure QD film.

Figure 8.1. Impact of Reduced Dimensionality (a) Charge carriers is restrained in a semiconductor (denoted by red) having a smaller bandgap than the cladding semiconductor (denoted by blue). The confinement alters the density of states

D(E). Ec is the conduction-band edge of a 3D semiconductor. (b) Threshold current-density development of heterostructure lasers. A noticeable decrease in threshold can be accomplished by introducing structures having reduced dimensionality: quantum wells and quantum dots.

Figure 8.2. (a) Representation of a mode-locked laser having hybrid mode locking. The structure of the ridge-waveguide laser is divided into gain section and absorber sections; these sections are separated by a small air gap. The asynchronous optical pulse train is generated by the modulated absorption (b) RMS timing jitter of a hybrid mode-locked QD laser being operated at 40 GHz, and derivative of the single-sideband noise integrated from 1 kHz to 1 GHz. (c) Left: External modulation of the optical pulse train of 40 GHz generates an optical data stream of 40 Gb/s. Right: Measurement of the autocorrelation of the 40 GHz pulse train, up-converted by two-stage optical-time-division multiplexing, with a pulse comb at 160 GHz.

Figure 8.3. Cross-gain modulation of a semiconductor optical amplifier (SOA) (a) Principle of Operation: a strong and modulated pump signal (Pin, denoted by red) impels the SOA into the saturation regime. As a result, the SOA gain is also modulated. The pattern of modulation can be shifted to a weak CW probe pulse at some other wavelength (Pin, denoted by blue), resulting into an inverted pattern of modulation on the probe signal in the output (Pout, indicated by blue). (b) The efficiency of the cross-gain modulation of a QD SOA (shown at top) and a conventional QW SOA (shown at bottom). The 3dB bandwidth of the QD SOA denoted by the horizontal line can be tuned at high injection current to beyond 40 GHz (c) Wavelength up-conversion by cross-gain modulation of a QD SOA and a pseudo-random RZ-OOK- modulation. Top figure: input eye illustration of an 80 Gb/s signal at a wavelength of 1292 nm, bottom figure: up-converted output signal at a wavelength of 1300 nm.

Figure 8.4. A vertical-cavity surface-emitting semiconductor laser (VCSEL) is shown (a) Cross section view of Schematic. The optically active medium containing the quantum dots (QDs) is clad with the help of 2 distributed-Bragg reflectors (DBR) creating an optical cavity. The radiation is emitted vertically (indicated by red arrow). (b) Top view of a particular device displaying the aperture and the contacts. (c) Emission spectrum displayed on a logarithmic scale. (d) Eye diagrams are presenting 20 Gb/s device's operation up to 120°C for large signal modulation. The open eyes show good signal-to-noise ratio at high operation temperature as well.

LIST OF TABLES

PREFACE

Presently, the primary energy resources of the world are fossil fuels in the form of coal, natural gas, and oil. Currently, the world is facing severe energy, environmental and economic crises due to higher energy consumptions coupled with depletion of fossil fuel stocks. It is essential to shift the environmental dynamics towards clean and sustainable energy sources to combat the ever-increasing environmental issues. Solar/Photovoltaic energy is available in great abundance on the earth. However, the efficiency and cost effectiveness of current photovoltaic devices presents huge challenges to realize the large-scale production of solar energy. However, modern photovoltaic materials based on nano-scale architectures offer excellent photovoltaic characteristics at nominal costs.

Development of novel photovoltaic devices with considerably enhanced power conversion efficiencies is essential for the conversion of solar energy (renewable green energy) to electrical energy. Development of nanostructured materials during the last 2 decades has provided a promising platform to achieve the green energy goals. Photovoltaic cells based on different nanomaterials have exhibited promising results in terms of efficiency and cost-effectiveness. Currently, thin film solar cells (Tandem Solar Cells), Quantum Dot Sensitized Solar Cells, and Multiple Exciton Generation based solar cells are being explored extensively. There is a range of books which thoroughly discuss the nanostructures based solar cells; however, the books covering the potentials of Quantum Dot Solar Cells (QDSC) are scarce. Therefore, this book is intended to provide a thorough insight into the history, modern trends and prospective technologies associated with QD photovoltaics.

The theme of the book is based on the contributions of material quantum dots in the photovoltaic industry. Nanostructured Quantum Dots are regarded as the magic materials when it comes to light-harvesting. The unique structure of quantum dots presents the opportunity to harvest maximum sunlight in a range of wavelengths due to their higher surface areas. Various photovoltaic cell materials and architectures are discussed in different chapters. Particularly, Chapter 1 presents the introduction and background for quantum dot based photovoltaics. Currently, silicon is One of the earliest photovoltaic materials. Silicon-based conventional solar cells are being used commercially. However, there is a scope to unearth the photovoltaic characteristics of silicon QDs. Chapter 2 focuses on novel photovoltaic characteristics of silicon quantum dots. Nowadays, various synthesis routes are being practiced to manufacture solar cells. Chapter 3 specifically discusses the fabrication and characterization approaches for quantum dots (QDs) and QD based photovoltaic devices.

Environmental sustainability is essential for the survival of life on earth. Modern photovoltaic devices are manufactured while considering the environmental concerns. Synthesis routes and characteristics of carbon quantum dots based green photovoltaic cells are discussed in Chapter 4. Solar efficiency is One of the major concerns in photovoltaic cells. The phenomenon of multiple exciton generations (MEG) for enhanced photovoltaic efficiency is demonstrated in Chapter 5. Carbon nanotubes (CNTs) and graphenes are regarded as the essential semiconducting materials due to their unique characteristics at nano-scale. Chapter 6 exclusively illustrates the photovoltaic behavior of graphene quantum dots. Chapters 7 and 8 focus on the challenges and prospective solutions provided by Quantum Dots for developing high-efficiency LEDs and solar cells.

This book is specifically important for QD photovoltaics because it encompasses all the essential information regarding the quantum dot based photovoltaic devices. This book is specifically designed for all the students, researchers and scientists out there looking to change the future of the energy world.

INTRODUCTION

Recently, humankind has learned how to obtain electricity from the solar energy system through phototermics and photovoltaic processes. The firsts operate under principles similar to those of conventional power thermal plants. The photovoltaic presents an important simplification compared to the usual energetic processes because they transform primary energy, the solar, into electricity in a direct way, that is, without intermediate transformations in other forms of energy. These devices are statics and not at all similar to conventional generators.

Processes that involve the dispersion or absorption of radiation doses are basically three, namely: the photoelectric effect (the process involved in the conversion of solar energy into electric energy), the Compton effect and the production of pairs.

The photovoltaic cell is a special design of the p-n junction or Schottky barrier diode. In a solar cell, incoming sunlight evicts electrons and their positively charged counterparts, called holes, in the middle layer of the device. The electrons then migrate to an electrode, releasing some electrical charge and return to the holes through another electrode to fill in the gaps.

It was Becquerel who first discovered that the light solar pruning can be converted directly into electricity in 1839 when I was observing the photovoltaic effect. So, in 1876, Adamas and Day found that selenium has photovoltaic properties. When Planck postulated nature of the light in the year 1900, the door was opened up for other scientists in building the theory.

Then in 1930, Wilson proposed the quantum theory of solids, which provided a theoretical link between the photon and the properties of the solids. Ten years later Mott and Schottky developed the theory of the solid state diode, and in 1949 Bardeen, Brattain and Shockley invented the bipolar transistor. This invention revolutionized the world of devices of solid state. The first solar cell was developed by Chapin, Fuller, Pearson, then in 1954, which had an efficiency of 6%.

The Sun provides the energy necessary to sustain the life in our Solar System. In 1 hour the Earth receives enough solar energy to satisfy the demand that is needed for the production of 1 year. In other words, this is about 5,000 times the energy that the Earth produces from all its energy sources.

The Sun is composed of a mixture of gases with a predominance of hydrogen. How the sun turns the sun hydrogen in helium in a thermonuclear fusion of massive reaction, the mass is converted into energy according to Einstein's famous formula $E = mc^2$. Because of this reaction, the sun surface is maintained at an approximate temperature of 5800 K. This energy is radiated from the Sunshine uniformly in all directions, from according to the black body radiation Planck's formula.

The optical absorption is responsible for the generation of the on of the electron-hole pair. When light strikes a material, it can be re-used or absorbed, transmitted, or absorbed. The absorption of the light is simply the conversion of the contained energy in the incident photon in some other form of energy. Some materials have the necessary properties to convert the energy of the incident photons in electric power.

When a photon is absorbed, it interacts with an atom delivering its power to an electron. This energy is governed by the rules of momentum conservation and energy conservation.

The photon of zero invariant mass has a momentum very small in comparison to the time of the electrons and holes. Depending on the energy of the photon, an electron can be elevated to a state of high energy inside an atom, or it can be to be released. The released electrons are then able to move through the crystal according to the present phenomena that can cause the electron to move, such as temperature, diffusion or an electric field.

Photons that are not absorbed cannot be to use to create useful energy. Only the amount of energy absorbed can be transformed into useful energy, the goal is to make the most of this fraction.

The wavelengths of visible photons are in the hundreds of nanometers (nm). At low energies (eV) typical for visible light, photons interact mainly with the valence electrons.

Semiconductors are solid covalent materials that can be considered as insulators, since their valence band is completely full, and the conduction band is completely empty at absolute zero. However, it has a range of forbidden energies between the conduction and valence bands of approximately 2 eV. For silicon(Si) the prohibited energy range is 1:14 eV.

An impurity that gives electrons is called donor impurity, and the resulting semiconductor is called type n (negative) because it has an excess of free electrons.

An impurity with electrons deficiency is called acceptant, and the resulting semiconductor is called the type p (positive).

The border between these 2 regions is known as the union p-n. The valence band represents the allowed energies of the valence electrons that are bound together to the atoms. The conduction band represents the permitted energies of the electrons that have been received the energy of some kind of mechanism and are no longer attached to specific atoms.

At T = 0 K, all allowed energy states in the valence band of a semiconductor are occupied by electrons, and there are no states allowed in the conduction band.

Since the driving process requires that the carriers pass from 1 state to another within an energy band, non-conduction can take place when all the states are occupied or when all the states are empty.

It's very important to remember that when you have a cell is illuminated, usually only converts 20% irradiance in electricity. When the cell is illuminated, the electron-hole pairs are produced by the interaction of incident photons in the cell. The electric field created by the union of cells produces a separation of the electron-hole pairs, with the electrons drifting in region n, and the holes drift in the region p.

CHAPTER 1

INTRODUCTION TO QUANTUM DOT PHOTOVOLTAICS

CONTENTS

1.1. BACKGROUND

Limited fossil fuel resources on the planet and increasing global energy demand necessitate the development of modern renewable energy sources. Amongst several energy technologies prevalent nowadays, photovoltaics technology is considered One of the economic and cleanest approaches to achieving the energy goals. However, tireless efforts are still required to make photovoltaics cost-effective over other contemporary technologies for energy generation (Service, 1996). The inclusion of nanoscale materials in photovoltaic applications has demonstrated different possibilities to manufacture low-cost (yet efficient) solar cells (Chiba et al., 2006). Recently, inorganic semiconducting nanocrystals (also called Quantum Dots or QDs) have been found to exhibit promising potential for next-generation solar cells (Robel et al., 2006; Kamat, 2007). Quantum dots have been investigated for photovoltaic applications because of their smaller size and composition dependent absorption characteristics (Vomeyer et al., 1994; Peng et al., 2000; Emin et al., 2009, 2010; Yu et al., 2003). Formerly unattainable with bulk semiconductor materials, the quantum dots permit energy level equivalence between potential donor and acceptor semiconductors, which is critical in designing cost-efficient photovoltaic devices (Gu¨nes et al., 2007a,b; Leschkies et al., 2007; Yun et al., 2009; Chang et al., 2010). Furthermore, QDs can be solution treated and could be an ideal alternative to universally used sensitizer molecules (Bang & Kamat, 2009).

Colloidal nanostructures of quantum-confined semiconductor materials are an emerging category of functional materials which are being developed for advanced solar energy technologies. One of the major losses in a thin film of the bulk solar cell takes place after a few picoseconds of the photon absorption, as photons having energies higher than the bandgap of the particular semiconductor generate charge-carriers containing surplus kinetic energy, which is eventually dissipated via emission of phonons. Nanostructures of semiconductors contain at least 1 dimension at particular nanoscale which is responsible for the production of quantum confinement effect. Nanostructured materials with quantum confinement phenomenon provide novel pathways for regulating energy flow for increasing the efficiency of the key photo-conversion step. This chapter focuses on the use of the modern research in the field of semiconducting colloidal quantum dots (nanostructured crystals confined in 3D) for the synthesis of cost-effective solar cells. Studies have revealed that these different nano quantum dot (QD) systems can bypass the single-junction Shockley-Queisser limit for conversion of a solar photon (Emin et al., 2010).

Nanomaterials constitute a flexible material platform which has extraordinary potential for providing new gateways for conversion of solar energy. The manufacturing, characterization, and use of these advanced nanostructures depend on the intersection of multidisciplinary fields, i.e., chemistry, physics, nanotechnology, materials science, etc.

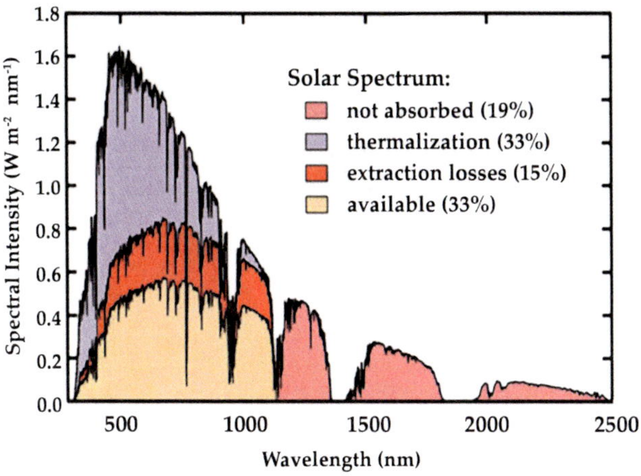

Figure 1.1. Spectral illustration of the silicon solar cell losses (bandgap = 1.1 eV). These losses are typically associated with the Shockley-Queisser limit and exemplify an upper limit for single-junction solar cells manufactured from bulk semiconductors. Thermalization signifies the biggest loss in this particular analysis which upsurges for the higher energy regions of the solar spectra [*Source*: https://www.sciencedirect.com/science/article/pii/S1369702112702201].

The chemical technologies are providing convenient and safe manufacturing routes based on solution phase synthesis which result in passivated monodispersed nanocrystals (NCs) of fine optoelectronic quality with an increasing level of control over shape, composition, and structure. These innovative structures offer new possibilities to physicists, material scientists, and engineers for controlling efficient flow of energy. One of the greatest scientific challenges related to solar energy conversion is the efficiency enhancement in the primary photo-conversion step.

In recent years, the phenomenon of multiple exciton generation or MEG has been studied extensively, where a photon containing at least double the energy of the semiconductor bandgap can generate multiple electron-hole pairs with a minimum waste of heat energy (Nozik et al., 2010; Beard, 2011).

1.2. THIRD GENERATION PHOTOVOLTAIC DEVICES WITH MULTIPLE EXCITON GENERATION

Conventional solar cells only extract a fixed amount of solar energy from a photon. However, the photons in the solar spectrum contain energies ranging from 0.4 eV to 4.0 eV (Figure 1.1).

The amount of the solar energy conversion to electrical energy mainly depends on the band-gap of the semiconductor, i.e., the photons having energy greater than the semiconductor bandgap loses its excess energy via thermalization (photonic emissions) while the low-energy photons are not absorbed in the semiconductor. Figure 1.1 illustrates that the amount of free energy in a typical single junction solar cell is around 33%, while an additional 33% is lost in the thermalization process. On the other hand, the remaining energy is distributed in unabsorbed photons and unavoidable thermodynamic losses.

These energy losses are related to the extraction of photo-excited electrons before radiative recombination at the contacts. The analysis displayed in Figure 1.1. 1 was initially presented by Shockley and Queisser in 1961 for a solar cell with a bandgap of 1.1 eV functioning under the global spectrum of AM1.5 with a power of 1000 W/m^2 (Shockley & Queisser, 1961). First and second generation of photovoltaic (PV) solar cells have excellent power conversion efficiencies (PCE) which are asymptotically reaching the Shockley-Queisser (SQ) limit. For instance, the record efficiency for GaAs based solar cells is presently 28.3%, while c-Si-based solar cells exhibit the efficiency up to 25% (Beard et al., 2012; Semonin et al., 2012).

The third generation of photovoltaics can have a higher conversion efficiency by circumventing One of the assumptions presented by SQ analysis. This practice can assist in recovering some of the lost energy or providing avenues for harvesting the energy of unabsorbed photons in a typical solar cell (Green, 2003). Apart from the efficiency considerations, third generation solar cells also account for economic considerations.

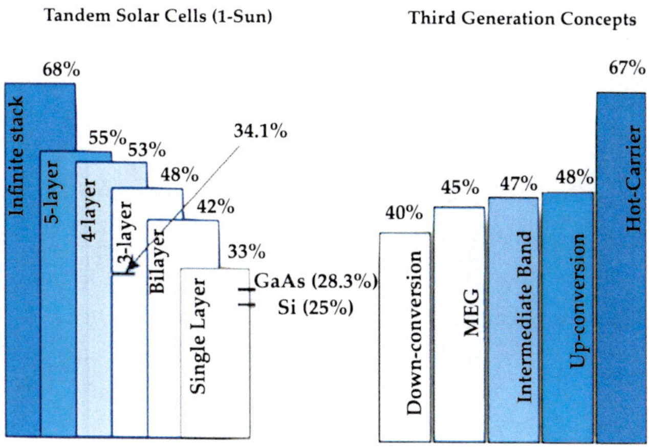

Figure 1.2. The overall balance-derived allowable power conversion efficiencies for a broad range of novel solar conversion designs. The original calculation for Schockley-Queisser limit of a conventional single-junction solar cell device is also displayed, as well as some record certified value solar cells with single and triple junctions [*Source:* https://www.sciencedirect.com/science/article/pii/S1369702112702201].

There are various approaches for solar conversion in third Generation Photovoltaics. Figure 1.2 illustrates the comparison of some important categories of solar cells. Semiconducting quantum-confined nanocrystals are being investigated as constituents in each of the above-shown approaches (Figure 1.2). The most traditional approach is based on a multi-junction or tandem solar cell which comprises of a pile of *p-n* junctions where the bandgap of the semiconductors decreases in successive layers. It is evident from Figure 1.2 that the theoretical energy conversion efficiency fluctuates concerning the number of semiconducting layers. Theoretically, an infinite number of p-n junctions can provide a solar efficiency of up to 68% under one-sun intensity (Green et al., 2015). Figure 1.2 displays the certified record power conversion efficiencies under 1-sun illumination for single junction GaAs and Si as well as a triple-junction cell (~34.1%). The PCE can rise beyond that of 1-sun illumination conditions under concentrated light, and the record efficiency for all solar cells is approximately 43.5% for 418-suns (Yin & Alivisatos, 2004; Yin et al., 2006). The material and manufacturing challenges offered by high-efficiency III-V multi-junction photovoltaic cells have led to their higher costs which limit their extensive utilization. However, low-cost multijunction solar cells based on quantum dots, organic absorbers, and their combinations are dynamically being investigated.

Other approaches for manufacturing of third generation photovoltaics include down- and up-conversion multiple exciton generations (MEG), hot carrier and intermediate band solar cells. This chapter discusses the utilization of unique characteristics of quantum dots to instigate MEG in solar cells. An ideal MEG absorber implicates the excitation of 2 electrons upon incidence of a single photon with energy equivalent to twice the bandgap of the material. The MEG approach is unable to extract all of the excess solar energy like an infinite tandem solar cell due to the presence of energy loss between multiple bandgaps starting at 1^{E}. On the other hand, the MEG approach is mainly attractive due to the capability of a single absorber layer to carryout exciton-generation. Conventional carrier contacts are typically used in MEG manufacturing, and the solar cell can be synthesized using solution processing methods.

1.3. COLLOIDAL QUANTUM DOTS

Quantum dots are typical 3-D confined nanocrystals of semiconducting materials. Quantum rods (QR) exhibit two-dimensional confinement while one-dimensional confined semiconductor nanocrystals are known as quantum plates, wells, or sheets. QRs and QDs are commonly manufactured under standard vacuum conditions typically conducive for batch reactions of organometallics in organic solvents (Murphy et al., 2006). The termination of crystals is usually carried out by organic ligands which maintain colloidal dispersion, stabilize crystal growth, and usually segregate them from their atmosphere. A broad range of different semiconducting quantum dots has been fabricated with dimensions ranging from 1 to 20 nanometers in length. A certain degree of parameter space is characteristic to the synthesis of quantum dot structures. Parameter space can be investigated and tailored to account for distinct control over architecture and composition.

Figure 1.3. TEM images of different sized of PbTe and PbSe QDs exhibiting the high crystallinity (insets represents small and wide-angle patterns of electron diffraction). (a-b) 9.0 nm QDs of PbTe (c) Cubic-like PbTe QDs of 14.3 nm length (d) 15 nm orderly array of PbSe QDs (e) QDs of PbSe (f) HRTEM of 27 nm cubics PbSe [*Source*: https://www.sciencedirect.com/science/article/pii/S1369702112702201].

The as-produced quantum dots are approximately all single crystal nanostructures suspended in a solution, and hardly exhibit twinning or other material defects. These QDs are faceted due to the retention of the underlying bulk material crystal structure. For instance, in Figure 1.3 shows TEM images of PbSe and PbTe QDs (Hughes et al., 2012). Regions (a) and (b) display smaller quantum dots which exhibit various facets symptomatic of the underlying crystal lattice of rock salt. In this particular case, the {111} and {100} facets are the most significant. The morphological changes can occur in QDs due to increase in their size by the addition of the monomers along different directions. Thus larger PbSe or PbTe QDs (regions (c-f)) display nanocrystal structure which has undergone faster growth in the <111> direction culminating in cubic nanocrystals with primarily {100} facets exposed. The morphology of the nanocrystal can impact the packing arrangement of QDs in films. Figures 1.4a and 1.4b exhibit hexagonal close packing while Figure (c-f) display cubic packing.

The shape of nanoparticles can be maneuvered by controlling the energetics of the distinct facets and the interaction dynamics of the ligands with different surface species. Figure 1.4 illustrates the TEM images of PbSe NCs developed under different reaction circumstances. In regions (a) and (b) the faces of {111} plane are blocked, and the growth of crystal is along the <100> direction generating hexapods (Schapotschnikow et al., 2010). Panel (c) demonstrates anisotropic NCs (e.g., QRs) of PbSe grown along the <100> direction through dipolar and ligand-ligand interactions (10). Control and manipulation of unexplored growth mechanisms for the production of a range of compositions and shapes is the contemporary research endeavor. Apart from the development of QRs, it is also assumed that platelets can be manufactured by the 2-D orientation of QDs into arrays which recrystallize as single crystal lattices (Schliehe et al., 2010).

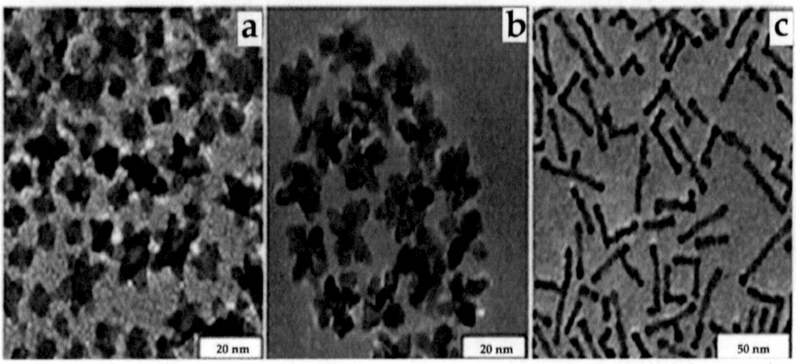

Figure 1.4. TEM images of quantum confined PbSe nanostructures exhibiting a few of the several shapes distinguishable by advanced synthesis methods [*Source*: https://www.sciencedirect.com/science/article/pii/S1369702112702201].

Quantum confinement effects tend to be critical when at-least 1 dimension of the nanocrystal approaches the length of de Broglie wavelength of holes and electrons in the bulk semiconductor, i.e., $\lambda_e = (h/m_{eff}kT)$, where m_{eff} represents the effective mass of the hole or electron for λ_h. For most of the semiconducting materials, λ_e and λ_h are in the range of 10 – 100 nm. For instance, λ_e = 24 nm for Si, and λ_e = 40 nm for PbSe. Another essential length scale in semiconductors is the exciton Bohr radius (a_B) that illustrates the distinctive segregation of a united electron-hole pair in a bulk semiconducting material. Excitons are quasi-particles comprised of correlated (United) electron-hole pairs which are mutually attracted due to the Coulomb potential. The exciton binding energy, for most of the bulk semiconductors, is small in contrast to kT; and hence, holes and electrons do not interact. On the contrary, the reduction of a semiconductor to dimensions smaller than a_B, holes, and electrons are confined to a specific volume where they continue to interact even at room temperature. The calculation of the Bohr radius (which typically fluctuates between 2 – 100 nm) can be made using the expression: $a_B = \varepsilon h^2/\mu e^2$ (where ε represents the semiconductor dielectric constant, and μ denotes the reduced operative mass of the hole and electron). For instance, a_B is 4nm for silicon and 7 nm for PbSe.

Figure 1.5 shows absorption spectra of PbSe quantum dots with an average radius which varies from 3 to 8 nm. The lowest feature of absorption blue-shifts concerning the reduction in diameter. The upper right graphical diagram illustrates the increased confinement in ground state wave-function which enlightens this shift pattern in the lowest-energy transitions. The decrease in the QDs size increases the kinetic energy and the separation distance between states. For example, PbSe NCs with 4.8 nm diameter exhibit an effective bandgap of around 0.82 eV, displaying a strong blue shift of > 500 meV induced by confinement, as compared to the bandgap of 0.28 eV for bulk PbSe. Variations of less than 1 atomic monolayers can have a significant impact on E_g as displayed for 4.8 and 4.6 nm QDs.

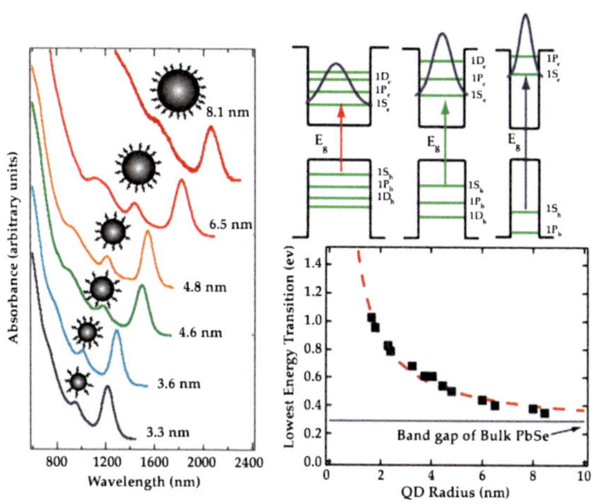

Figure 1.5. Absorption spectra of Lead Selenide quantum dots ranging from 3.3 nm to 8.1 nm exhibit strong quantum-confined absorption with 1st exciton shift. Various discrete transitions are apparent and signify discrete exciton transitions. The bottom right graph illustrates that QDs Eg for PbSe varies nearly as 1/r. Top right graphs demonstrate the enhanced quantum confinement experienced by 1st exciton due to the decrement in particle size [*Source*: https://www.sciencedirect.com/science/article/pii/S1369702112702201].

Quantum dots can exhibit features distinctive of both small semiconductors and large atoms; hence, the language used for the description of their photophysical characteristics is, usually, a blend of semiconducting and atomic language. For instance, the LUMO and HOMO are categorized by 2 quantum numbers (i.e., *n* & *L*) while ignoring spin-orbit interactions for angular momentum and the energy levels. HUMO and LUMO are typically labeled according to the values of angular momentum as P (L=1), D (L=2), S (L=0), etc., while the LUMO-HUMO separation is commonly termed as the bandgap, E_g.

Speculative and investigational study of semiconducting nanostructures is significantly expanding our knowledge of size-dependent electronic structures, interfaces, sizes, composition and the influence of surface properties on energy flow and carrier dynamics. The discussion on the enhancement of multiple exciton generations using quantum-confinement is of vital importance (Shabaev et al., 2006; Isborn et al., 2008; Rabbani & Baer, 2010). Crystal momentum typically defines the energy dispersion relationships in bulk semiconducting materials and governs selection rules, but it exhibits relaxed state in nanostructured semiconductors. In impact ionization (which is the bulk equivalent of MEG) both energy and crystal momentum need to be conserved during the ionization process. On the other hand, only energy conservation is required in QDs, growing the number of ways for MEG phenomenon. Additionally, MEG is instigated by the Coulomb interactions which is supported by quantum confinement (Lin et al., 2011).

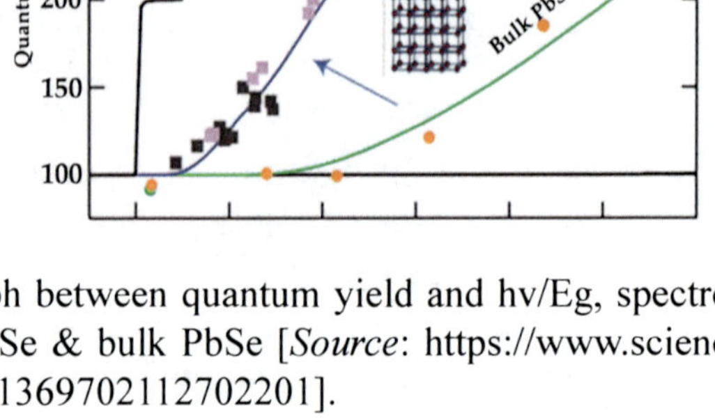

Figure 1.6. Graph between quantum yield and hv/Eg, spectroscopically measured for QD PbSe & bulk PbSe [*Source*: https://www.sciencedirect.com/science/article/pii/S1369702112702201].

It has been established via ultrafast spectroscopic characterizations that MEG is roughly 2–3 times more efficient in QDs of PbSe as compared to its bulk counterpart. Figure 1.6 displays the plot for some excitons generated per photon (quantum yield) concerning $h\nu/E_g$ for bulk semiconducting PbSe compared with exciton generation rates for nanostructured QDs of PbSe. The quantity $h\nu/E_g$ signifies the number of available bandgap energy provided by a single excitation photon. The lines demonstrate a model which describes the MEG efficiency. Quantum rods (QRs) exhibit further improved quantum yields (Cunningham et al., 2011). Current research endeavors must address the approaches for further improvements in the MEG process to reach approach the fundamental limit as close as possible.

1.4. QUANTUM DOT SOLAR CELLS

Colloidal quantum dots based solar cells have been the focus of extensive research during the last few years. As a potential alternative to conventional silicon solar cells, QD solar cells are the most suitable choice for next generation of advanced photovoltaic devices. Colloidal QDs are fascinating in photovoltaics research because of their convenient solution processability which is advantageous for their amalgamation into several solar cells (Kongkanand et al., 2008; Lee et al., 2008a). A detailed review of the modern developments in various QD solar cells synthesized from colloidal QDs has been presented in the latter sections of this chapter.

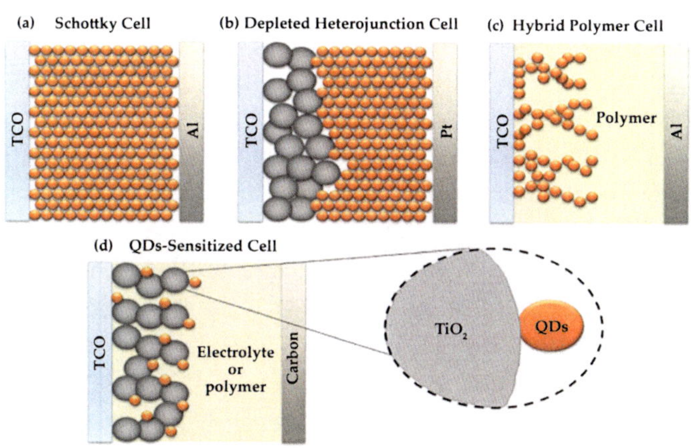

Figure 1.7. Comparison of different QDs based solar cells. (a) Schottky cell; (b) A typically depleted heterojunction solar cell; (c) hybrid polymeric solar cell; d) QD sensitized solar cell. [*Source*: https://www.sciencedirect.com/science/article/pii/S1369702112702201].

The chapter also discusses various fabrication techniques, working principles, benefits and drawbacks of different QD based devices (Pattantyus-Abraham et al., 2010). Major topics of discussion include integration of QDs in Schottky solar cells, hybrid inorganic-organic solar cells, depleted heterojunction cells, bulk heterojunction photovoltaic cells, quantum dot sensitized photovoltaic cells and extremely thin absorber photovoltaic cells (Koleilat et al., 2008; Johnston et al., 2008;).

Figure 1.7 shows typical device architecture employed in QD solar cells. The interest in quantum dot solar cells is further pronounced by a phenomenon known as a multi-exciton generation (MEG) which can be utilized for developing high-efficiency photovoltaic cells (Johnson, 2008). Quantum dot photovoltaics present the prospects of enhancing the energy conversion efficiency of a cell beyond the conventional Schockley and Queisser limit of around 32% for silicon-based solar cells (Shockley & Queisser, 1961). In recent times, efficient MEG phenomena have been reported in PbS, PbSe, PbTe, CdSe and Si QDs (Murphy et al., 2006; Schaller et al., 2006; Beard et al., 2007; 2010; Pijpers et al., 2009; Sukhovatkin et al., 2009). However, the realization of this phenomenon has not been fully explored for QD based solar cells. Fundamental studies are required for the development of high-efficiency QD solar cells with ultrafast exciton dynamics (Schaller et al., 2004).

In the last few years, numerous review articles have been prepared on QD solar cells which include bulk heterojunction cells and QD sensitized solar cells. Since substantial research has been focused on QD sensitized solar cells, several review articles can be found on this subject concerning various aspects of QD sensitized solar cells. One of the initial reviews on QD sensitized solar cells had been presented by Kamat (2008). Furthermore, Hodes (2008) demonstrated a comparison based study on the working concepts of QD sensitized solar cells and dye-sensitized solar cells. Similarly, a review paper on QD sensitized solar cells has been published by Ru¨hle et al. (2010) highlighting the part of molecular dipoles in quantum dot's energy-level alignment. Colloidal QDs have also been employed in bulk heterojunction photovoltaic cells. Kumar and Scholes (2008) and Arici and Sariciftci (2004) have reviewed the latest developments in this field. Owing to the significant research interests concerning 1D structures, Yun et al. (2009) demonstrated the integration of colloidal nanoparticles (QDs) in nanowires and nanorods based solar cells. Additionally, Talapin et al. (2010) have studied a broad range of QDs applications in advanced optoelectronic devices which include photodetectors, light emitting diodes, and quantum dot solar cells.

The concept of utilizing colloidal quantum dots in photovoltaic cells has been investigated for around 2 decades. Nozik (2002) proposed 3 common strategies to integrate solution grown quantum dots into solar cells. These strategies are discussed in the following.

- Use the quantum dot to sensitize semiconductors with wide-bandgap.
- Position the QDs in close contact with hole/electron conducting polymers.
- Constitute QD arrays where the quantum dots are electronically united to permit for efficient hole/electron conductivity (Nozik, 2002).

The previous research on QDs involved the direct development of quantum-confined sensitizers using porous semiconductors with a wide-bandgap analogous to a dye-sensitized solar cell. At that time, colloidal synthesis of quantum dots was still in its initial stages, but the broadband characteristics of absorption were more beneficial than those of some dye molecules (Murray et al., 1993; Colvin et al., 1994). The progression of synthetic techniques towards monodispersed QDs grown from a solution and the concepts of solution cast QD layers combined with organic polymers have resulted in promising charge transfer between quantum dots and conducting polymers (Greenham et al., 1996; Ginger & Greenham, 1999). Low carrier mobility characteristics in QD based films, typically required close association with a charge- splitting interface for operation of QD photovoltaic cells. Considerable efficiencies have been achieved by synthetically stretching the colloidal particles to improve electron transport through the combined QD/polymer film (Huynh et al., 2002). This concept has been significantly upgraded to achieve an efficiency of up to 3.13% with CdSe nanocrystals (Dayal et al., 2009) (Figure 1.8)

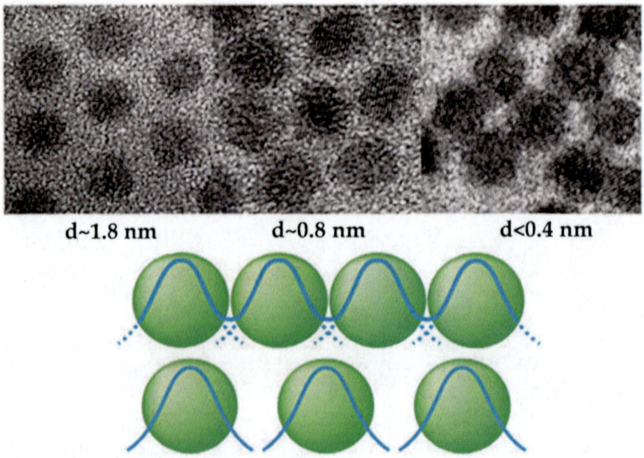

Figure 1.8. The selection of QD surface ligand controls the QD-QD distance. As produced films having oleic acid capping group reveal an average QD-QD space of about 1.8 nm while ethylenediamine treatment can result in the reduced distance, i.e., up to < 0.4 nm. The immediate advantage of a reduced QD-QD distance is better QD-QD electronic interaction. This phenomenon is graphically illustrated in the lower panel which shows the overlapping electronic wave-functions for excited electrons between adjoining QDs. [*Source*: https://www.sciencedirect.com/science/article/pii/S1369702112702201].

Each of the concepts discussed above relies on the intimate contacts of quantum dots with other materials to attain charge-transport and charge-separation, where quantum dots serve primarily as the light-absorbing constituent. The third approach for QD solar cell requires an extremely coupled QD-array which performs the jobs of light absorption, electron-hole splitting, and successive charge-carrier transport. On the contrary, insulating aliphatic ligands on the surfaces of QDs prevent efficient transport of charge, although photocurrent spectrum on an all quantum dot arrays with TOPO ligands has been recorded by researchers in 2000 with 190 nanometer thick films of CdSe quantum dots inserted between ITO and several metals (Ginger & Greenham, 2000). Guyot Sionnest (2003) and co-workers established a plain ligand exchange design whereby a solid quantum dot film is primarily formed from the as-made quantum dots followed by soaking in a solution of a different shorter ligand (Yu et al., 2003). The conductivity of the resultant QD film increases considerably as the longer insulating ligands are replaced by shorter ligands. Arango et al. (2009) revealed that shorter ligands (e.g., butylamine) lead to amplified photoconductivity of CdSe based QD films, and they also demonstrated that film treatment in NaOH could remove the ligands (Jarosz et al., 2004). Talapin and Murray (2005) developed field effect transistors (FETs) using PbSe QDs in conjunction with hydrazine (i.e., the shortest possible diamine) to replace or remove the oleate ligands, thereby impressively enhancing electrical coupling (Figure 1.9)

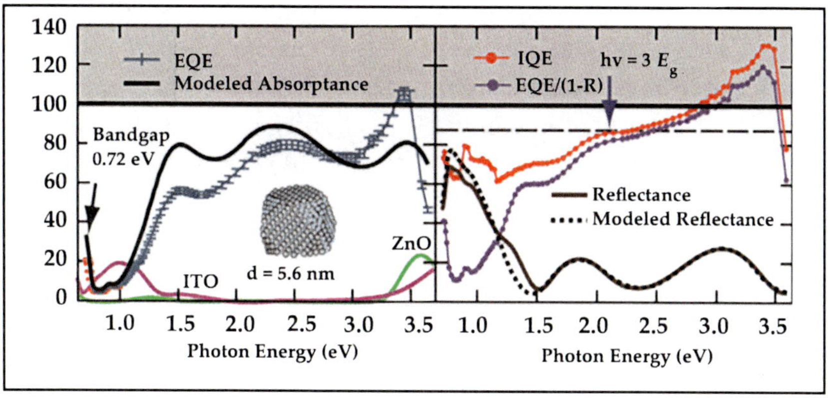

Figure 1.9. Spectral characterization of QD based PbSe solar cell displaying MEG. The left panel shows a plot of external quantum efficiency (EQE, represented by blue curve), and the modeled absorbance of the active cell layer (black), ITO (fuchsia) and ZnO (green). The right panel showed a graph with modeled (dashed black) and measured (brown) reflectance, the internal quantum efficiency (IQE, represented by orange) and the EQE regulated by reflectance (purple) [*Source*: https://www.sciencedirect.com/science/article/pii/S1369702112702201].

Exchange of ligands with hydrazine results in significant volume loss, pinholes and cracks are filled by progressive treatment and deposition which leads to the layer-by-layer deposition and removal of ligands create smooth pinhole free conformal films having electronically coupled quantum dot (Klem et al., 2007; Beard et al., 2010; Luther et al., 2008). All-QD PbSe and PbS Schottky junction photovoltaic cells have originated from these synthesis concepts and these quantum dot layers have been united with transparent oxide semiconductors to enhance the charge carrier collection and photovoltage (Clifford et al., 2007; Luther et al., 2008; Leschkies et al., 2009).

The optoelectronic characteristics of QDs mainly depend on a subtle interaction between film morphology, nanocrystal surface chemistry, and inter-particle interactions. The prime function of the ligand in a quantum dot is mere to define the distance between different QDs. Studies have shown that the change in ligands results in variation of inter-QD distance, i.e., substituting the oleic acid ligands with those of aniline or ethylene diamine results in the QD-QD distance which fluctuates between 1.8 nm and < 0.4 nm (Choi et al., 2009). The instant output is to couple the QDs electronically so that the films show a drastic decrease in resistivity (Murphy et al., 2006).

However, the exchange of ligands has more potential than controlling the inter-QDs movement. While working on the impact of hydrazine treatment on QD films of PbSe, Talapin, and Murray (2005) found that the arrays with *n*-type behavior could be converted to *p*-type by exposure of the films to mild hearing or vacuum. Recently, scientists utilized halide ions for surface passivation. It was observed that passivation resulted in the generation of *n*-type QD films PbS (Li et al., 2016). On the contrary, conventional thiol treatments generate *p*-type behavior (Liu et al., 2008, 2009, 2010). Recently, Tang et al. (2009) built a photovoltaic cell which involves the combination of passivation strategies to develop homo-junction PbS quantum dot solar cells with substantial efficiencies analogous to the current record of 7% for PbS QD devices (Ip et al., 2012). Intentional doping is carried out for bulk semiconductor materials to provide functionality to the characteristics of semiconductors. A similar level of control over doping strategies is still under investigation for quantum dot thin films. Exchange of ligand has been investigated with hydrazine, amines, carboxylic acids, thiols, and several inorganic ligands (Zarghami et al., 2010). For photovoltaic applications halogens, thiols, or blends of passivates have demonstrated effectiveness, and they persist as a promising pathway. This photovoltaic behavior is quite likely due to the curvature of quantum dots which presents an intricate surface structure with several steric restrictions (Talapin et al., 2009).

Apart from the synthesis routes for QD films, the thorough understanding of defect states, carrier-lifetime and mobility is essential for developing functional optoelectronic devices with higher efficiencies (Hyun et al., 2008; Tang et al., 2010; 2011; Semonin et al., 2011). Presently, novel characterization techniques are being extensively employed to unearth the numerous phenomena taking place in the QD world. The investigation of mid-gap surface states has been carried out with time-resolved infrared spectroscopy, photoluminescence, transient photovoltage spectroscopy and optical field-effect transistors (Chappel et al., 2011; Galland et al., 2011). Terahertz and microwave probes can provide important information for controlling and understanding carrier transport and carrier lifetime in QD films.

The dynamics of the multiple exciton generation processes in QD films are estimated to be considerably different from those observed for colloidal QD solutions. The introduction of new pathways has resulted in improved photovoltaic characteristics of QDs present in different forms. The electronic coupling of close-packed QDs instigates the movement of excited carriers nearby QDs followed by their tramping in new surface states induced by the chemical treatments. These phenomena can have a significant impact on yields and spectroscopic measurements of MEG. A few spectroscopic techniques using the microwave and near-infrared probes have proposed that it is likely to sustain MEG in electronically coupled QDs films. However, the best way to determine the effect of quantum dot coupling on MEG involves the application of spectrally resolved photocurrent techniques in an operational photovoltaic device (Torrisi et al., 2012; Nagpal & Klimov, 2011).

1.5. MEG SOLAR CELLS

Tandem architectures and MEG are 2 of the third-generation photovoltaic approaches. Functioning of quantum dot solar cells has been demonstrated in the previous section of this chapter. In the case of tandem solar cells, the "smoking gun" is an illustration of an open-circuit voltage which exceeds the open circuit voltages of the subcells present in the device and is roughly equal to the sum of the sub-cells. This phenomenon has been illustrated in various research articles related to tandem QD solar architectures. A tandem solar cell with dual-junction architecture comprises of a small bandgap device on the bottom followed by a larger bandgap device present on top of the smaller bandgap device. The sub-cells in a tandem solar cell are linked in series; therefore, the overall device current will be controlled by the sub-cell which generates the least amount of current. Tandem architectures also have some additional complications, i.e., the top cell typically absorbs a substantial portion of the solar spectrum which could, otherwise, have been absorbed by the bottom cell (Luther et al., 2010).

The intricacy of transparent interlayers, current matching absorbers, and multiple junctions is eradicated in a MEG device. The suitable bandgap range for using MEG in a photovoltaic cell is from 0.6 to 1.1 eV, as compared to conventional photovoltaic devices where the optimal range is 1.1–1.4 eV. The demonstration of MEG behavior in an efficient QD photovoltaic cell has been made in the recent studies by various scientists. These research studies mainly focus on determination of the ratio of electrons created per photon absorbed by the photovoltaic cell. This ratio is termed as the internal quantum efficiency (IQE), which demonstrates that a measurement value above unity would exemplify the generation of more mobile carriers than the number of photons absorbed. However, the IQE is difficult to measure practically; therefore, the number of electrons generated. As a result, *incident* photons on the solar cell are more commonly measured (known as external quantum efficiency or EQE). This quantity is typically lower than the IQE (Figure 1.10)

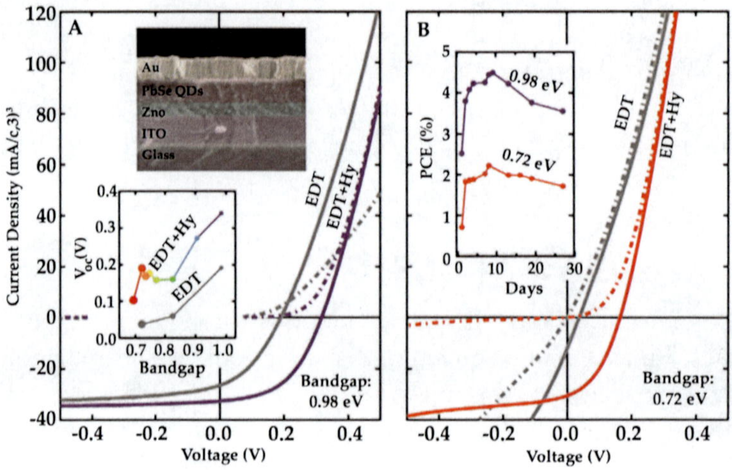

Figure 1.10. Current-voltage (JV) characteristics of PbSe quantum dot solar cells displaying MEG phenomenon. JV plots are obtained under illumination power of 100 mW/cm2 for 2 bandgaps of QDs, 0.72 eV (orange curve, i.e., B) and 0.98 eV (purple curve, i.e., A) and), in contrast with the conventional single treatment, EDT (i.e., grey curves). Dark JV curves having no illumination are graphed in dot-dashed curves. An SEM cross-section of the device architecture is displayed in panel A, an evaluation of voltage with hydrazine+EDT against EDT-only is displayed in the inset to A, and the aging behavior of power conversion efficiency (PCE) is exhibited in the inset to B. [*Source*: https://www.sciencedirect.com/science/article/pii/S1369702112702201]

A typical photovoltaic cell usually shows an EQE peak of 106% (which can further exhibit increment up to 114% by incorporation of an anti-reflection film), confirming that a single photon can generate multiple excited electrons which can be gathered at the device electrodes. The utilization of an optical model depicted an IQE of around 130% for the same wavelengths. Essentially, the extracted IQE was consistent with previous spectroscopic analyses of MEG. This demonstration employed a multiple-passivation approach which involved layer-by-layer deposition of PbSe quantum dots with ethane-dithiol (EDT) ligand followed by deposition with hydrazine ligand. This disparity is visible in the current-voltage characteristics of the EDT+hydrazine and EDT-only approach. Ip et al., (2012) have demonstrated that the amalgamation of multiple treatments is a potential pathway for the synthesis of better-passivated quantum dot films. This kind of extensive development of quantum dot films as a functional material for photovoltaic devices will be critical for the development of *any* next-generation photovoltaic cells.

1.6. QUANTUM DOT SENSITIZED PHOTOVOLTAIC CELLS

1.6.1. Working Mechanism

The previous research on quantum dots sensitized solar cells, or QDSSCs has been illustrated by Weller's group and Vogel et al. (1990). These types of functional photovoltaic cells are an alternative to DSSCs (dye-sensitized solar cells) (Mishra et al., 2009). The difference between DSSCs and QDSSCs lies in the sensitizers used to extract the energy from sunlight. The sensitizers in a DSSC device is typically an organic dye or metalorganic complex while inorganic semiconductor QDs based sensitizer is used in QDSSCs. Figure 1.11 illustrates a schematic for QDSSC and its operating principle. The brief working principle of QDSSCs is described in the following text.

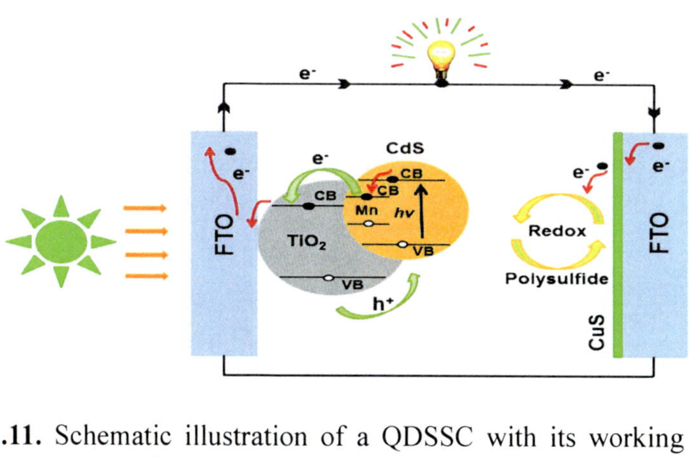

Figure 1.11. Schematic illustration of a QDSSC with its working principle [*Source*: http://pubs.rsc.org/en/content/articlelanding/2015/dt/c4dt03063j/unauth#!divAbstract]

Upon solar illumination of the cell, the quantum dot sensitizer linked to the TiO^2 (titanium dioxide) film absorbs incident light. This light absorption instigates the electron excitation in quantum dots from the valence bands to the conduction bands. In this phenomenon, each of the excited electrons generates a positive hole along with it. The electrostatically united electron-hole pair is known as exciton gets dissociated at the TiO^2/QDs interface. The excited electrons are inserted into the nanoporous TiO_2 electrode, culminating in the oxidation of the quantum dots photosensitizer. The injected electrons are then transported via an external load to a counter electrode. On the other hand, the oxidized quantum dots accept the electrons from a redox mediator restoring the ground states of atoms in QD sensitizer. Subsequently, the oxidized mediator enters to the counter electrode to get reduced. In this pattern, 1 cycle is accomplished, and the gained power is because of the absorbed photons by nanocrystals of semiconductor material (Sun et al., 2010; Shen et al., 2010).

1.6.2. QD Sensitization of TiO_2 Electrodes

Metallic oxides of wide band-gap semiconductor materials such as ZnO and TiO_2 have been extensively used as porous electrodes in QDSSCs. Morphological characteristics of these metallic oxides have been dynamically explored to achieve excellent power conversion efficiencies for QDSSCs developed from these oxides (Chen et al., 2006; Toyoda et al., 2007; Yum et al., 2007; Lee et al., 2008b; Baker and Kamat, 2009; Liu & Kortshagen, 2010). For instance, metallic oxide nanowires have been used for increasing the conductivity of thin films (Chen et al., 2009; 2010a). Likewise, nanotubes of metal oxides have been extensively used to improve the carrier transport and realize light trapping effects. Inverse opal architectures have also been used for enhanced light trapping and carrier transports. Figure 1.12 displays typical morphological features of ZnO and TiO_2 films utilized for the synthesis of QDSSCs. Metal oxide sensitization using nanocrystals focuses on following approaches:

- QDs are grown directly onto the electrodes (TiO_2) using chemical bath deposition (CBD) technology (Toyoda et al., 2007).
- SILAR (successive ion-layer adsorption & reaction) procedure is carried out (Chang & Lee, 2007; Lee et al., 2009b, 2010).
- Bifunctional linkers are responsible for linkage of QDs to metal oxide surfaces (Robel et al., 2006; Sambur et al., 2010; Sambur & Parkinson, 2010).

- Physisorption process is employed for the association of QDs with metal oxides (Blackburn et al., 2005; Giménez et al., 2009).

Chemical bath deposition process is One of the primary techniques typically employed for deposition of semiconductor quantum dots on the surface of metallic oxide (Šimurda et al., 2006). This technique gained reputation due to its convenient process and its ability to produce components on a large scale. However, in spite of its simplicity in general, the process of chemical bath deposition is unable to account for precise control over spectral properties and the particle size of QDs. In recent times, chemical bath deposition technique has evolved into a more lucrative technique known as SILAR. Concurrently, another method which also gained wide appreciation is the linker assisted self-assembly technique. In this method, pre-synthesized QD nanocrystals are adsorbed on the surface of metal oxides by using molecular linkers having several functional groups (Chen et al., 2011).

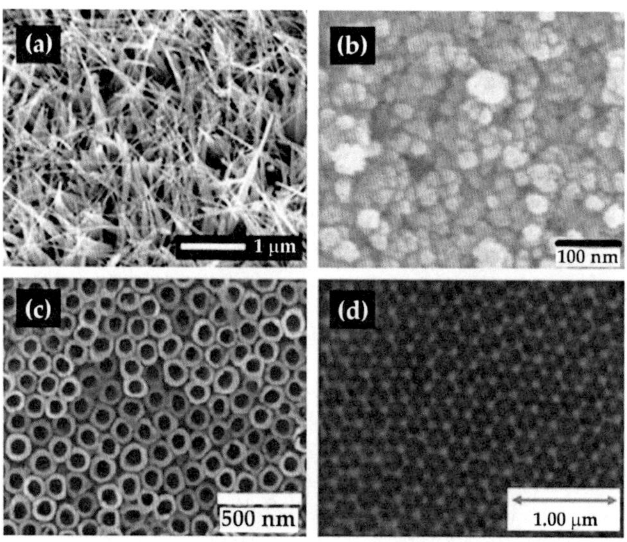

Figure 1.12. SEM images of ZnO and TiO2 films with different morphological features: (a) nanowire structures; (b) mesoporous structures; (c) semiconducting nanotubes; (d) inverse opal. (Toyoda et al., 2007; Lee et al., 2008b; Chen et al., 2010a) [*Source*: https://pubs.acs.org/doi/abs/10.1021/acs.chemrev.5b00063]

Linker assisted self-assembly method enables accurate control over the spectral characteristics of QDs and the interaction between the substrate surface and QDs while ensuring coverage of monolayer surface. Another method which gained enormous attention during the last few years is typically based on physisorption principle. This method involves the sensitization of metallic oxides through physisorption. Metal oxide electrodes are usually dipped into QD solutions with the immersion time approaching 100 h. The QD sensitized electrodes are integrated into photovoltaic cells for obtaining the better results which are superior to those achieved with bi-functional linkers.

1.7. FUTURE CHALLENGES AND DIRECTIONS

Exceeding the Schottky-Quisser limit for single junction photovoltaic cells is both a technological and scientific challenge, and the utilization of semiconductor nanocrystals to improve the primary photo-conversion procedure is a promising pathway towards achieving such goals. The results of MEG solar cells are extraordinary not only as a convincing demonstration of MEG phenomenon but also as an illustration that the 'extra' charge carriers can be gathered in an appropriate quantum dot photovoltaic cell. Therefore, One of the limitations of the SQ limit (i.e., high-energy photons can only generate 1 electron-hole pair) can be surpassed. However, the current MEG photovoltaic devices offer a yield of about 4% for photocurrent collection from multiple excited charge carriers per photon. The present challenge to further enhance the efficiency of MEG, as well as to carry out the improvements in the fundamental architectures of QD films. 1 possibility of proposed research is to investigate MEG in a range of compositions, structures, and shapes. Quantum platelets (QPs) having one-dimensional confinement and quantum rods or quantum wires with 2-D confinement are both comparatively unexplored, and quantum rods have already offered some promising results which exhibit further improvement of MEG (Beard et al., 2009; Talgom et al., 2011).

Other possible approaches to high-efficiency photovoltaic devices, especially multi-junction photovoltaic cells, are very favorable avenues as well. However, the effectiveness of these approaches requires the provision of fundamentally well-controlled semiconducting materials. Kramer and Sargent (2011) proposed that carrier mobility, trap level position and trap density are useful parameters for investigating the efficiency of a cell. Moreover, the carrier lifetime (intrinsically associated with trap density) is an available and essential parameter to explore in this endeavor. Lifetime and mobility determine the diffusion length of a particular semiconductor, which is a critical parameter for determination of efficient cell-thickness before the dominance of recombination reactions. This length, currently about 100 nm, need to be extended to the length regions of optical and near Infrared absorption range (Choi et al., 2011; Wang et al., 2011).

Quantum dots of lead chalcogenide may be regarded as an interesting test-bed to comprehend the mechanisms of incorporating quantum-confined semiconductor materials into appropriate photovoltaic architectures. However, lead chalcogenide QDs may also exhibit potential as a long-term semiconducting material for large-scale positioning. Recent studies have illustrated that PbS QD photovoltaic cells can show stability for more than 1000 hours under constant illumination. This particular study advocates that lead chalcogenide photovoltaic has the potential to exhibit excellent characteristics in the 3 essential economic considerations: stability, cost, and efficiency. Wadia et al. (2009) conducted an analysis study concerning the natural abundance of raw materials for photovoltaic devices and found that PbS ranks positively regarding economic considerations. However, PbS is usually overlooked due to its low bandgap in the bulk state (57). On the contrary, the toxicity of lead also plays an important role in the selection of materials for photovoltaic applications. However, the precautions can be made to manufacture lead sulfide-based QD solar cells.

PbS QDs have to potential to provide ample lead to synthesize roughly 1000 m^2 of photovoltaic cells (supposing a 1 μm QDs layer), which could deliver 100 kW peak power with 10% efficiency. This energy would be enough to run a common electric car continuously after making considerations for average insolation at latitudes. Nevertheless, quantum-confinement opens the pathways to the novel and unexplored horizons of low bandgap semiconducting materials which have not previously received attention for photovoltaic applications. Moreover, novel photophysics like MEG offers new ways to utilize the available solar energy in a more fruitful way (Nalwa, 2004; Kim et al., 2008).

CHAPTER 2

SILICON QUANTUM DOTS FOR PHOTOVOLTAIC APPLICATIONS

CONTENTS

2.1. INTRODUCTION

Currently, the world makes use of energy at a rate of about 15 trillion watts (TW). By 2050, this demand is anticipated to become more than double, even with aggressive conservation and energy proficiency measures (Kramer, 2012; Sargent, 2012; Graetzel et al., 2012). Continual improvements in existing energy networks will not be enough to supply this demand in a sustainable manner. High pressure on prevailing networks is bound to raise energy prices; this together with the hazard of global warming has raised a pervasive interest in renewable energy. The challenge in embracing the photovoltaic technology as the prime source of power generation on a worldwide scale is intensely decreasing the price/watt of delivered solar electricity—by roughly a factor of 5–10 to compete with nuclear and fossil electricity and by a factor of 25–50 to keep pace with primary fossil energy. Worldwide scientists are working hard to bring these costs down. This can be achieved in 2 ways; decreasing the manufacturing cost of materials and improving the efficacy of the devices (Gao et al., 2014; Ye et al., 2017; Rauf & Rezai, 2017; Figure 2.1).

Commercial silicon solar cells have been used extensively in the last decade. However, presently, Quantum dot (QD) sensitized solar cells are getting considerable attention due to their potential for replacing the current conventional solar cells as highly efficient and low-cost alternatives. Control of shape and size in materials at nanoscale allows the researchers to tailor the physical properties of different materials. Nanomaterials presents novel prospects in engineering, renewable energies, microelectronics, and life-science. Conversion of solar light into electrical energy is considered as a critical factor for continuous provision of electrical energy in future. However, the costs involved in contemporary solar power energy systems are quite high. The efficiency of the solar cells also needs to be improved as compared to the efficiency offered by conventional solar cells, i.e., below 20%. The above-mentioned factors strongly suggest the exploration of modern nanostructures for developing cost-efficient photovoltaic cells. Presently, the interest in investigating the properties of silicon-based nanostructures in photovoltaic cells has enormously increased. Silicon has already proved itself as an excellent semiconductor material owing to its increased use in high-speed transistors. This chapter summarizes the most significant characteristics of Si-QDs (silicon quantum dots) for photovoltaic applications.

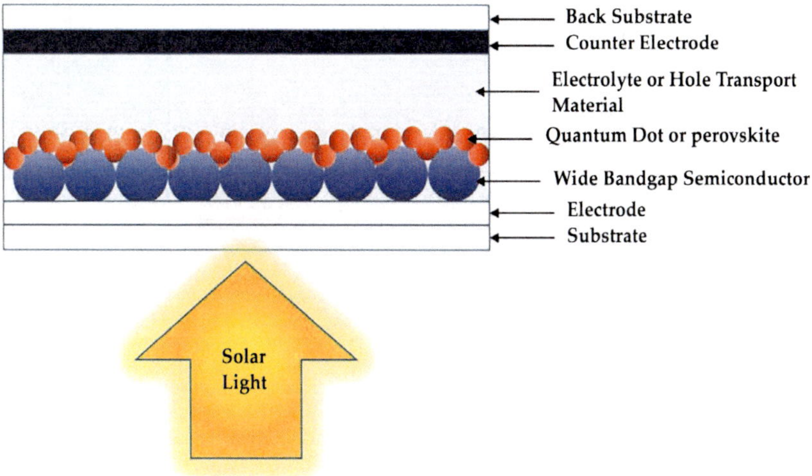

Figure 2.1. Schematics of a typical perovskite or a quantum dot sensitized solar cell.[*Source*: https://www.sciencedirect.com/science/article/pii/S1364032117301399].

Various attempts have been made to comprehend the dependence of silicon semiconductor sensitized solar cell material properties which involve conventional "One variable at a time (OVAT)" methodology for experimentation (Mora-Seró & Bisquert, 2010; González-Pedro et al., 2010). Even though, based on the OVAT, some qualitative suggestions of the combined effects and significance of optimizing the combination of multiple variables have been made in the literature, still, no quantitative data has been found on the relative significance of each of the aforementioned factors as well as their combined impacts (interactions) on the performance of solar cells (Etgar et al., 2012; Etgar, 2013; Kriegel et al., 2012).

Statistical analysis of studied experimental data can play a key role in the transformation of the experimental data into critical information for the development of fundamental science (Genovese et al., 2011; Hetsch et al., 2011; Zhao & Burda, 2012). The current drastic capital cuts for basic research due to recession have instigated a discussion among the research and the academic community to develop as well as utilize methodologies for more rapidly advancement of systems knowledge at nominal costs (Maathuis et al., 2009; 2010; Tsujimoto et al., 2012; Rhee et al., 2013). Various approaches have been applied in energy sciences. Nevertheless, these approaches have never been formalized with the academic framework for analyses of historical data in solar cell materials. In this chapter, a unique methodology has been presented to provide deep insight into silicon-based quantum dot devices (Panjikar et al., 2009; Ehrlich & Rarey, 2011).

2.2. SILICON QUANTUM DOTS

Si-QDs vary significantly in a series of properties from bulk silicon:

- Si-QDs exhibits an increase in the energy gap with a reduction in the dimensions of the dot. Hence, by controlling the size, it is possible to shift the material's energy bandgap to the optimal energy gap (i.e., 1.6 eV) for supreme efficiency of solo junction solar cells, or to stack cells with altered bandgap and therefore different spectral response together for forming tandem cells (Blow, 2009; Aliberti et al., 2010; Shrestha et al., 2010).
- Si-QDs exhibits strong photoluminescence in the visible/ visible-red region of the solar spectrum. Further, in literature, external quantum efficiencies as high as 60% has been reported. This type of strong photoluminescence can be employed in photoluminescence downshifters for increasing the efficiency (Murray et al., 1993; Alivisatos, 1996; Al-Ibrahim et al., 2005).
- Since in quantum dots, the distance between the energy states is wide; it leads to slowing down the relaxation of excited carriers concerning bulk silicon. If the carriers can be extracted from the dot prior relaxing to the edge of the band gap, we could increase the open circuit voltage of the solar cells. Previously, this idea has been explored in the models for hot carrier solar cells (Amato & Brunetto, 1996; Amato et al., 1998; Sensfuss et al., 2005).
- Another phenomenon has been investigated for use in solar cells which is the generation of many excitons in silicon quantum dots. An exciton having energy much greater than the bandgap (at least double), has an increased chance to split into 'n' excitons instead of losing the surplus energy in the form of heat and comforting to the edge of the bandgap. These many excitons, if successfully extracted from the dot, would result in an increased current of the solar cell. Else, if multiple excitons get recombine in a radiative manner, the material then would exhibit an improved photoluminescence quantum efficiency (Balberg et al., 2007; Anopchenko et al., 2009; Marconi et al., 2009).

The chapter can be organized in the following manner: in the 1st section, a summary on the properties of Si-QDs has been presented focusing particularly on the properties significant for usage in solar cells such as photoluminescence properties, band-gap opening, electrical transport, etc. Significant research has been carried out in this field ever since the observation of strong photoluminescence from porous silicon which took place in 1990 and a series of reviews and excellent books are available for the reader interested in them (Canham, 1990; Amato et al., 1997; Cullis et al., 1997) (Figure 2.2).

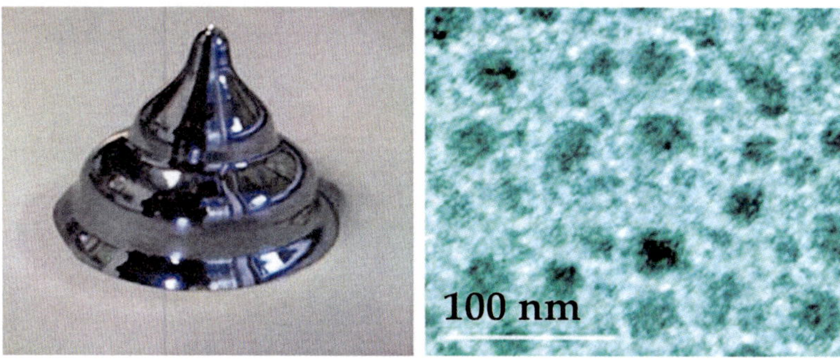

Figure 2.2. The upper portion of the single crystalline silicon wafer (left). In the photograph, the diameter of the wafer is 100 mm. The largest silicon ingots which have been grown by the Czrochalski method have a length of ~2m and diameter around 30 mm. HRTEM photograph of tiny silicon quantum dots in a SiO_2 matrix (Right). Si-dots look like dark structures, and the (111) lattice plans can be seen. The measurement of the Si-quantum dots presented in the image is between 2 and 5 nm (Bisi et al., 2000; Brabec et al., 2011). [*Source*: https://www.intechopen.com/books/quantum-dots-a-variety-of-new-applications/silicon-quantum-dots-for-photovoltaics].

In the 2nd part, we have presented a series of ideas and concepts, which focus on the realization of solar cells grounded on silicon quantum dots. Few of these concepts have already been demonstrated successfully with another type of semiconductors. For instance, tandem cells that consist of stacked solar cells with each of the cell reaping a different portion of the solar spectrum, however, they rely on somewhat rare elements such as In, Ga, or As and hence will not play a major role in generating TW of electrical energy from sunlight. Additional concepts like hot-carrier cells are very encouraging regarding their extraordinary thermodynamic efficiency limit (~85%), however, so far no hot carrier-solar cell has been fabricated. As a result, experimental research in this field focuses on the study of hot carriers in various quantum confined and bulk semiconductors as well as on the development of discerning energy contacts for carrier extraction (Becquerel, 1867; Berberian-Santos et al., 2005; Antonova et al., 2008).

2.3. PROPERTIES OF SILICON QUANTUM DOTS

2.3.1. Optical Properties

In the figure above, a piece of a single silicon crystal ingot having a diameter of 10 cm is shown. Further, a high-resolution TEM photograph of minute silicon nanocrystals inserted in SiO_2 having dimensions around 2–4 nm have also been shown. Although the properties of crystalline silicon can be considered homogeneous for dimensions which range from some tenths of nanometers up to several centimeters, the characteristics of Si-QDs are:

- mainly dependent on size,
- subjective to the properties of the interface between matrix and QDs,
- In materials, the electrical transport depends largely on the distance between the QDs.

For a thorough understanding of the properties of Si-QDs, it is beneficial to recall some of the fundamental characteristics of silicon. Silicon forms a diamond structure in the crystalline phase. In this structure, every atom of silicon is surrounded by 4 nearest neighboring atoms which form a tetrahedron. The silicon's unit cell consists of 8 atoms and size of the lattice constant "a" is 0.543 nm. Owing to the minor distance of the atoms in the solid state form, the energy levels of the atoms overlap and results in the formation of bands. The outmost electrons from 2 bands which are referred to the valence band and the conduction band (Bruchez et al., 1998; Vinciguerra et al., 2000). The valence band is entirely filled with electrons at zero degrees Kelvin, whereas, the conduction band is utterly empty of electrons. A prohibited gap (band gap) separates the conduction band from the valence band. In case of silicon, at room temperature, this bandgap is around 1.12 eV (or ~1107 nm). Silicon is an indirect band gap semiconductor, which implies that the top of the conductance band and the bottom of the valence band are not aligned in-k space. The following Figure 2.3 shows the electronic band structure of silicon centered on k· p calculations (Brongersma et al., 2000; Klauk, 2012).

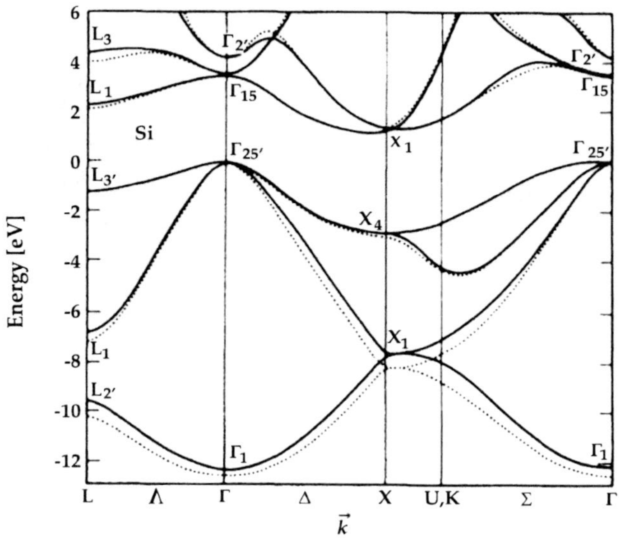

Figure 2.3. The figure shows general band structures of Si. For silicon, results are shown for local (shown by dashed line) and nonlocal (indicated by solid line) pseudo-potential calculations (Chelikowsky & Cohen, 1976). [*Source*: https://www.intechopen.com/books/quantum-dots-a-variety-of-new-applications/silicon-quantum-dots-for-photovoltaics].

The maximum of the valence band is present at the Brillouin zone center, whereas, the minimum of the conduction band is near to the point X (where k is not 0) at the Brillouin zone boundary. As optical transitions preserve both energies as well as momentum, hence, the excitation of an electron from the topmost level of the valence band to the bottommost level of the conduction band can take place only through the support of another process to preserve the momentum (Chan & Nie, 1998; Buhbut et al., 2010). In case of pure silicon, this takes place through the transferal of the electron momentum to a phonon (therefore, the transition is called indirect, and the absorption close to the edge of the band is very low that can be nicely observed from the absorption coefficient k in the Figure 2.4 (left). For the similar reason, the likelihood of relaxation of an electron that is excited from the conduction band to the valence band escorted by the natural emission of a photon is low. At higher energies, the absorption increases because of the beginning of the lowest energy direct transitions (Shalom et al., 2010; Pucker et al., 2012; Lee et al., 2013). At 359 nm, the absorption reaches a first maximum assigned to the E_1 transition, whereas, it reaches a strong maximum at 296 nm owing to the E_0 transition and further another weaker maximum at 225 nm. Figure 2.4 (right) displays the absorption edge of silicon. Here, the structures can be recognized as excitonic transitions. Due to the interaction in between the excited electron of the conduction band and the extra electrons (hole in the valence band), an exciton is formed. Owing to the complicated nature of the edge, excitonic transitions also become partially allowed through emission or absorption of photons and the associated transitions can be seen as fine structures in the absorption spectrum (Shaklee & Nahory, 1970; Chaouiet al., 2008). In bulk silicon, the exciton binding energy is comparatively small (~15 meV). In case, we decrease the dimensions of a semiconductor such as silicon in one, 2 or 3 dimensions on nanometric scale, the material properties will be affected by the spatial confinement, that is, the energy states of holes and the electrons lose their continuity and resultantly become discrete (Chapin et al., 1954; Chelikowsky & Cohen, 1976). Consequently, the emission and absorption of light when holes or electrons move from 1 state to another get affected. Often, nanostructures are classified according to the number of dimensions in which electrons (carriers) are at liberty to move in 0-D (quantum dots, nanocrystals, porous silicon), 1-D (quantum pillars and wires), and 2-D (quantum wells) (Linnros et al., 1999).

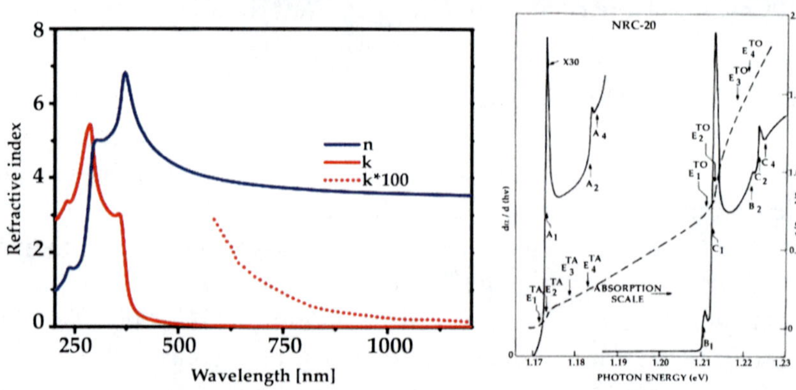

Figure 2.4. The experimental dielectric function of Si (red imaginary part (or absorption coefficient), blue – real part of the refractive index, (left). The absorption edge of c-Si measured at 1.8 K from (Shaklee & Nahory, 1970). [*Source*: https://www.intechopen.com/books/quantum-dots-a-variety-of-new-applications/silicon-quantum-dots-for-photovoltaics].

In the subsequent sections, we will concentrate mostly on results achieved on 0-D silicon quantum structures, even though silicon quantum wells and quantum-wires are very alike in their properties. Moreover, a series of encouraging and interesting results to be used in solar cells were particularly obtained on silicon quantum wires (Tsakalakos et al., 2007; Kelzenberg et al., 2010). One of the important effects of the confinement is its impact on the electronic properties of the system particularly on the energy of the band gap (Chuah et al., 2010; Zemnukhova et al., 2012).

The above figure shows an increase in the energy gap for attenuating size of silicon quantum structures (Lockwood et al., 1992). It is clear from the figure that a considerable effect on the energy gap is witnessed for Si-QDs that are smaller than 5 nm and for dots having a size of 2 nm, the gap is expected to be as big as ~2 eV. Experimentally, the shifts of the valence band and conduction bands can be observed in x-ray absorption methods (Lockwood et al., 1996). In addition to varying bandgap energy, the quantum confinement also affects the transition possibilities. Owing to the localization of the excitons (carriers) in a very insignificant volume, the selection rules of k are relaxed, and thus the probabilities of transition are increased (quasi-direct transitions). Most of the initial work on silicon nanocrystals was implemented on porous silicon. Porous silicon was discovered more than 50 years back (Uhlir, 1956) and is normally obtained via electrochemical etching of bulk silicon with HF. The range of permeability can be in dimensions from nanometers up to micrometers and under perfect etching conditions, we can obtain a spongy material having wall dimensions of a few nanometers. Intensely visible photoluminescence experimented from porous silicon in 1990 (Canham, 1990) and the luminescence was attributed to quantum confinement effects (Cullis & Canham, 1991). Intense research on porous silicon was carried out in the subsequent years for its application in silicon photonics and sensing. This research has been reviewed in a series of books (Amato et al., 1997; Bisi et al., 2000; Canham, 2007; Feng & Tsu, 1994). Silicon that is highly porous has a huge surface area, which is accessible for interacting with the ambient either through physisorption or chemical reactions of molecules. For instance, at room temperature, oxidation of porous silicon is likely to continue for weeks given that the surface is not accurately passivated. Additionally, because of the porosity, the material is also mechanically fragile and unstable. Due to this reason, the scientific community has been looking for alternative methods for preparation of better passivated as well as mechanically more stable Si-QDs (Daldosso et al., 2003; Arnberg et al., 2012). Currently, there exists a wide spectrum of fabrication techniques which permits to form Si-QDs in a solid-state phase, in the gas phase or solution. A common thing in most of the fabrication methods is that the silicon quantum dots exhibit a certain variation in shape and size (Dovrat et al., 2004; Dietmueller et al., 2009). As an example, in Figure 2.4, we can observe the absorption coefficient as well as the photoluminescence spectra of silicon quantum-dots gained by thermal reinforcement of a silicon-rich silicon oxide film (excess of silicon or silicon rich means that the film has abundant amount of silicon with respect to the proportion of 1: 2 among Si and oxygen in stoichiometric silicon dioxide). Due to the sizes of the light spot and the large density of silicon quantum dots in these types of measurements, a vast number of quantum dots can be measured at the same time (about ~1012 Quantum-dots in the ellipsometric measurements and ~1011 quantum-dots in the photoluminescence). The acute points of energy transitions, which have been well defined in the absorption spectrum of c-Si (for instance, the E1 transition at 3.395 eV), are extended and poorly defined owing to this averaging over a huge number of marginally different silicon quantum dots (Del Canizo et al., 2009; Di et al., 2010). By increasing silicon content, the absorption coefficient increases. The photoluminescence spectra which has been measured at room temperature of these samples exhibit a wide photoluminescence band in the VISNIR region, which transitions from 800 to 850 nm for declining silicon content. This sort of behavior has been reported by several studies, and only a few of them are cited in this chapter (Rinnert & Vergnat, 2003; Umezu et al., 2000; Yang et al., 1997; Pucker et al., 2000) (Figure 2.5).

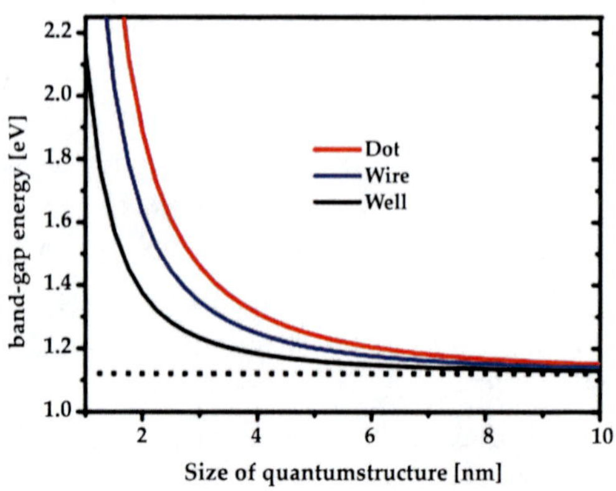

Figure 2.5. The optical gap in silicon quantum wires, wells, and dots, against systems diameter. The transition energy has been calculated for the lowest electron, and heavy hole-energies have been calculated for infinite confining potentials (Lockwood et al., 1992). [*Source*: https://www.intechopen.com/books/quantum-dots-a-variety-of-new-applications/silicon-quantum-dots-for-photovoltaics].

Further insight into the structure of Si quantum dots can be obtained by analyzing the photoluminescence (PL) decay curves. Comparable to what has been found in porous silicon, the PL decay exhibits a dispersion in decay life period, not solely a function of emission wavelength, but likewise for a fixed wavelength (Sinke et al., 2007; Feltrin & Freundlich, 2008). These decay curves are subjective to the dispersion in shape and size of the nanocrystals which results in different radiative transition possibilities and by trapping and migration of excitons. At room temperature, decay times are normally in the range of 10 to 300 μs (Wilson et al., 1993). Further, the decay time rises to milliseconds for low temperatures (such as <20 K) (Dovrat et al., 2004; Brongersma et al., 2000; Vinciguerra et al., 2000). The dependence of the photoluminescence decay curves on temperature can be explained by attributing the luminescence to the recombination of powerfully localized excitons in the Si quantum dot. Owing to the exchange interaction between the hole and the electron, the exciton levels are fragmented by energy. The upper state is a single state, whereas, the lower level corresponds to a triplet state (Wolfsegger et al., 2007; Wadia et al., 2009; Wan et al., 2011). The ground state of the exciton is a singlet state, and hence the transition to the triplet state is not allowed (the probability of non-zero transition is owing to spin-orbit interaction mixing the singlet and triplet states marginally), consequently, at low temperatures, the radiative decay rate is smaller (Forrest et al., 1982; Franzo et al., 2002). At elevated temperatures, the singlet state gets populated and the rate of decay increases (Figure 2.6).

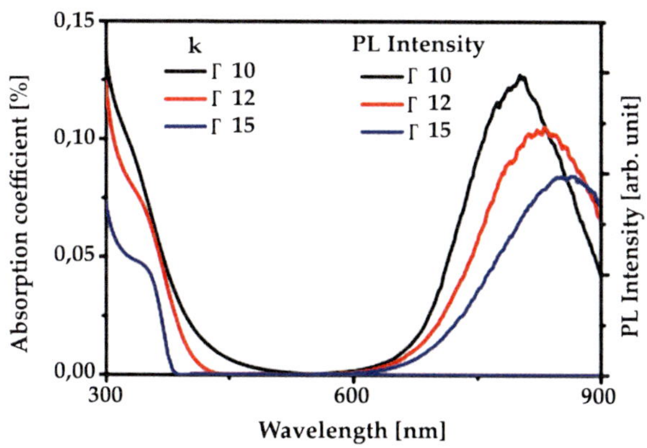

Figure 2.6. RT photoluminescence spectra and absorption coefficient of films having Si-QDs developed from thermal galvanizing of silicon-containing oxide films [*Source*: https://www.intechopen.com/books/quantum-dots-a-variety-of-new-applications/silicon-quantum-dots-for-photovoltaics].

2.3.2. Electrical Transport Properties

Initial work on charging in Si-QDs implanted in an oxide matrix and conduction processes was carried out back in 1984 by DiMaria. This work was performed some years before the observation of light emission from porous silicon which stimulated the boom of research in the field of silicon photonics (DiMaria et al., 1984a; DiMaria et al., 1984b). DiMaria and her team studied the electro-optical and the electrical properties of annealed silicon-rich-oxide carrying Si-quantum dots and observed electroluminescence; after this study and observation, they suggested carrier relaxation in quantum restricted silicon as the origin and talked about the transport of carriers in the material. The mechanism of conduction in dielectric matrices carrying Si-QDs depends largely on the type of matrix (silicon-nitride, silicon dioxide, and silicon carbide are among the most commons one) as well as on the distance between the Si-QDs. When preparation of Si-QDs is carried out by phase separation from silicon-rich oxide, two different regions can be observed in conductivity measurements: first, a region in which the amount of surplus silicon is so large that the Si-QDs do not remain isolated anymore but end up in forming a percolation backbone; second, a lower Si-surplus region in which the Si-QDs are isolated and conductivity is dominated by tunneling practices (Balberg et al., 2007). In the section having large Si-excess, the conductivity is normally in the order of 10^{-7} (Ohm-cm)$^{-1}$, which is almost three orders of magnitude lesser as compared to the conductivity of high resistivity floating zone silicon (which is ~10^{4} (Ohm-cm)$^{-1}$). In case of films that are doped heavily with boron, conductivity values around 0.1 (Ohm-cm)$^{-1}$ were reported by Mirabella et al. (2010). Nevertheless, as can be seen in (Balberg et al., 2007), the photoluminescence which was observed for isolated Si-QDs is absent in this high Si-excess regime (higher than the percolation threshold) due to the enlargement of the cluster size and therefore relaxation of the quantum confinement effect. Due to this reason, these clustered structures appear to be less favorable for some of the illustrated application of Si-QDs that have been reviewed here. Conductivity in SiO_2 films having isolated Si-QDs is in the magnitude of 10–10 (Ohm-cm)$^{-1}$ (Balberg et al., 2007). In previous years, an intensive investigation was carried out on films that represent this type of structure, particularly for an understanding of silicon-based light emitting devices (Perálvarez et al., 2006; Franzò et al., 2002; Prezioso et al., 2008; 2009). For the realization of light emitting diodes, we replace the oxide in a metal-oxide-semiconductor (MOS) structure by the Si-QDs carrying material, and the thickness of the top electrode is reduced to acquire a semitransparent electrode. Under forward bias condition, this diode shows electroluminescence in the spectral region near 730830 nm (1.5–1.7 eV). Forward bias depends strongly on the composition of the active material. In case of LED's, the Si-QD needs to be excited by the carriers that are injected from the external circuit. The energy gap of Silicon dioxide is around ~9 eV, and the barrier heights are approximately 4.7 eV and 3.2 eV for holes and electrons respectively (Prokes et al., 1998; Pucker et al., 2000). Since there is the difference in the height of the barrier, hence, it is much easier in case of a SiO_2 matrix to inject electrons rather than holes. Direct tunneling of carriers via an oxide is the prominent conduction process only if the thickness of SiO_2 is lesser than ~1.5 nm. For greater distances, transport is estimated to take place via Fowler Nordheim tunneling, in which the barrier height is decreased by the application of electric field and the electrons are accelerated (Kovalev et al., 1998; Klampaftis et al., 2009). After that, these hot electrons can end up in the formation of exciton-hole pairs through impact ionization of the nanocrystals, which afterward relax in a radiative manner. The significant information for using Si-QDs in solar cells as derived from these studies is twofold:

- Conductivity is slightly low.
- It is not easy to have a balanced situation regarding hole and electron current (at least if silicon dioxide is the embedding material).

Currently, some strategies are under investigation for improvement of carrier transport both in and from quantum-confined Si-QDs. By accurately placing the Si-QDs and controlling the distance among the dots in a range of 1.5–2 nm, it is possible to obtain carrier injection through direct tunneling of both electrons and holes and in structures like these, electroluminescence was observed by applying extremely low voltages (Anopchenko et al., 2009). A standard approach for fabrication of such multi-layer stacks with controlled Si-quantum dot distances has been explained in subsequent sections. An alternative method to improve the conductivity is to replace silicon dioxide with some other materials having smaller band-gap than silicon. Some of the widely investigated matrices include SiC and Si_3N_4 with band gaps of 2.85 and 5.3 eV, respectively. For Si_3N_4, the conduction band offset is 2.4 eV, whereas, in case of SiC, it is 2.95 eV. Recently, conductivities in the range of 10^{-4} (Ohm-cm)$^{-1}$ to 10^{-2} (Ohm-cm)$^{-1}$ have been reported for Si quantum dots in silicon nitride matrix (Wan et al., 2011), which signifies a huge increase about the values of conductivity observed for Si-QDs in SiO_2.

Lastly, the conductivity of materials containing Si-QD can be increased by doping. Normally, SiQDs in silicon dioxide matrix is doped with either phosphor or boron to obtain n-type and p-type conducting films, which would permit to engineer p-n junctions just like the ones for standard silicon solar cells. Large concentrations of dopants are required for doping of Si-QDs. The typical concentrations of dopant are approximately 1 atom %. A rise of several orders of magnitude in the conductivity has been reported for both phosphor and boron doping. Si-QDs can be doped with phosphorus or boron, which appears to act as electron donor or acceptor similar as in bulk-silicon. It is interesting to note that phosphor and boron doping has opposite effects on the efficiency of photoluminescence (Hao et al., 2009). In case of phosphor doped films, the intensity of photoluminescence increases in doped films, whereas, in case of boron doped films, it decreases. Present studies specify that phosphor doping decreases the number of dangling bonds at the quantum-dot/SiO_2 interface (Stegner et al., 2008), whereas, boron doping consequences in an opposing manner by increasing non-radiative recombination centers (Green et al., 2010; Zatryb et al., 2010; 2011).

2.4. SI-QUANTUM DOTS IN LATEST SOLAR CELLS

Presently, 3rd-generation solar cells (as solar cells based on quantum-dots) are targeting considerable increases in efficiency of energy conversion. Keeping in viewpoint, the employment of the energy confinement of silicon-based quantum dot nanostructures, presently, there are various concepts of solar cells that can be found in the literature. With the help of these concepts, four possible approaches can be retained:

- Down-shifting and down-conversion of solar cells (Klampaftis et al., 2009; Richards, 2006a): In this case, the target is to make a more proficient consumption of the short wavelength fraction of the solar spectrum, chiefly by making multiple low-energy photons from the absorption of single high-energy photon (for which the efficiency of cell quantum (QE) is low), or else by shifting photons with high energy to a lower energy photon for which the quantum efficiency (QE) of cell is high.
- Hot carrier solar cells (Luque & Marti, 2010, Nozik et al., 2010): in this case, energy conversion from solar-to-electricity is enhanced by decreasing energy losses associated with the absorption of

solar photons with energy greater than the band gap of the active photovoltaic substance.

- Multi-junction or all-silicon tandem solar cells (Conibeer et al., 2008; Di et al., 2010): In this method, the anticipated increase in efficiency is owing to the multi-bandgap approach. The cell having the highest band gap is positioned on the very top, whereas, the cell having the lowest band gap is placed at the very bottom of the tandem stack. Each of the cells absorbs the amount of light which it can convert effectively, with the rest of light passing through to the underlying cells.
- Organic/Inorganic solar cells (Buhbut et al., 2010): In this approach, the principle scopes of the usage of QDs is twofold: First, the light harvesting can be enhanced by broadening the area in which light is absorbed by using QDs of different size; second, the lifespan of the cell is improved by replacement of organic molecules with more stable inorganic modules (Min et al., 1996; Edmonds et al., 1994).

2.4.1. Si- Quantum Dots as Spectral Converters for Solar Cells

The main objective of spectral converters is to utilize the short-wavelength (λ) part of the solar spectrum more efficiently. Under a spectral converter, we are aware of a film or a material, which through absorption and re-emission alters the spectral distribution and amount of photons of the solar spectrum that are impinging on the converting layer. In this context, 2 distinct physical mechanisms have been explored: down-conversion and down-shifting. In the down-conversion mechanism, 1 photon with a certain energy results in 2 photons (in the ideal scenario) having energies. The performance of solar cells improves by down-conversion which minimizes the thermalization losses in solar cells (Nesheva et al., 2002; Mocatta et al., 2011). The maximum improvement in efficiency for a solar cell having a bandgap of 1.1 eV (crystalline silicon) was reported to be 38.6% positioning the spectral converter on the top of the solar cell (Trupke et al., 2002). Down-conversion is primarily observed for rare-earth ion doped materials and takes place either on single ions or else involves energy transfer processes between different rare earth ions. A description of rare-earth based down-converting phosphors can be found in (Richards B.S., 2006b). Till now no down-conversion process has been reported for Si-QDs. Still, the generation of multiple excitons (MEG) has been reported for Si-QDs, which in some way is similar to down-conversion as it results in forming more than 1 low energy exciton from a high energy exciton. The following sections will discuss MEG in the context of Si-QDs (Ruan et al., 2003; Van Sark et al., 2004). In the process of down-shifting, a high energy photon is first absorbed by the spectral converter and then re-emitted at lower energy. Figure 2.7 depicts the process of down-shifting in silicon QDs together with a radiative transfer of energy from the QDs to the solar cell. Incident high energy photons absorption and their re-emission at lower energy can be acquired with every three-level system.

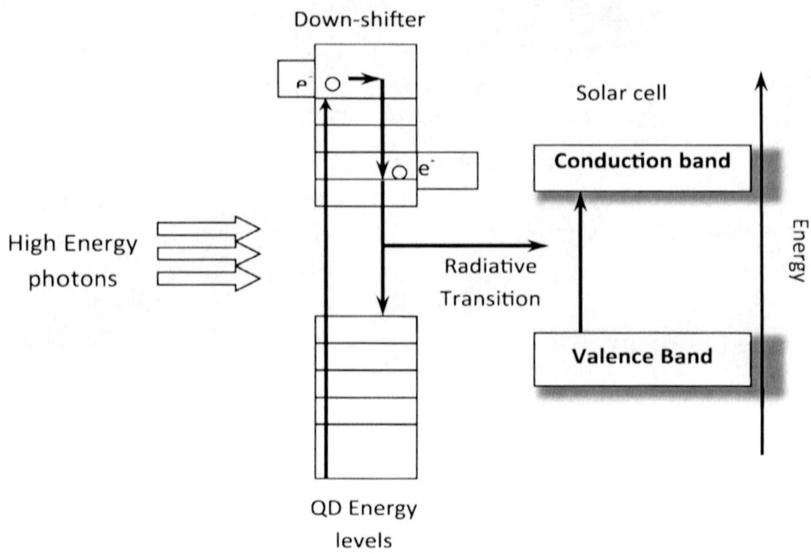

Figure 2.7. Schematic illustration of a solar cell energy diagram in conjunction with a spectral down-shifter [Source: https://www.intechopen.com/books/quantum-dots-a-variety-of-new-applications/silicon-quantum-dots-for-photovoltaics].

Fast nonradiative relaxation takes place amid the intermediate level and the highest excited level (Shockley & Queisser, 1961). In this scenario, the discharge of only a single lower energy photon is escorted by the radiative recombination of the electron initiating from the intermediate level to the lowest level. As an outcome of the luminescent process, a fraction of the incident photons gets shifted to longer wavelengths before reaching the active photovoltaic material of the device. The main purpose of the down-shifting practice is to maximize the number of photons incoming in the spectral region where there is maximum internal quantum efficiency (IQE) of the solar cell (O'regan & Grätzel, 1991). In solar cells that are based on crystalline silicon, the IQE is lesser for photons having wavelengths shorter than ~450 nm (because of recombination of carriers generated near the surface – known as surface recombination); whereas, it is very high between ~500 and 1000 nm. Due to this reason, downshifting of UV-blue photons to the red-NIR section can lead to a larger short-circuit current. The two other electrical features of the device, i.e., the fill factor and the open-circuit voltage will not change drastically, as the resistance of the device or the electronic characteristics of the semiconducting material remains unaltered.

A symbolic schematic representation of a photovoltaic spectral converter has been illustrated in the Figure 2.8. It comprises of four separate layers (Pavesi et al., 2000; Park et al., 2009). In this scheme, Silicon QDs, i.e., the active material is positioned on the front surface of the device. Further, the layer is isolated electronically from the solar cell through an insulator layer which means that the coupling between the solar cell and the active medium is purely radiative. On the rear surface of the device, a perfect mirror is located to supply high internal surface reflectance for every angle of incidence of light. Silicon QDs needs to take care of the following trends to be employed as down-converter for all kinds of existing solar cells:

- The absorption band must be large enough to cover the area where there is low External Quantum Efficiency of the cell (UV-blue section in Silicon-based solar cells);
- The absorption coefficient should be as high as achievable;
- The emission band should cover a spectral band where there is best external quantum efficiency of the cell (red area in Silicon-based solar cells);
- The energy difference in between the emission band and absorption band, i.e., the difference between the valence and conduction band must be large enough to evade phenomenon of re-absorption of the emitted photons.

One of the significant benefits of employing quantum dots, in general, is that both light emission and absorption can be tuned by selection of the dot diameter. This also holds true in case of Silicon QDs, as down-shifters in solar cells. Additionally, the large quantum efficiency of Si-QDs (over 50%, Walters, 2006), is another significant requisite for enhancing the efficiency of the solar cell. Initial simulation results completed on ideal QDs by van Sark have shown the great capability of QDs as down-shifting species (van Sark et al., 2004). The calculation has proved that quantum dots with emission around ~600 nm upsurges the multi-crystalline solar cell short-circuit current by almost 10%. The results of simulation indicate that the most advantageous effects will be acquired if the solar cell's spectral response has a particular form in such manner that the spectral response at low wavelengths is low; whereas, it is high at high wavelengths. While taking into consideration the critical parameters which makes an impact on the performance of down-conversion solar cells, three requisite factors can be pinned down:

- The distribution of energy of the incident spectra.
- Second is the presence or absence of an anti-reflecting coating on the forward-facing of the solar cell.
- The third is the surface recombination.

Jestin et al. (2010) have demonstrated through his experiments that the impact of a down-conversion layer carrying Silicon QDs on the Power Conversion Efficiency (PCE) is somewhat low (Hong et al., 2010).

Further, both simulation, as well as experimental results, have verified that we can observe an increase in the performance of the cell only in case of cell's poor performance in the UV-blue region (for instance, very high surface recombination velocity). In this scenario, solar cells having a luminescent down conversion layer can leave behind normal c-Si-based solar cells. Svreck et al. performed the first experimental tests using silicon QDs as a down-converter (Svreck et al., 2004) and described the embedding of Silicon QDs into a solution based on spin-on-glass anti-reflecting SiO_2. In this scenario, the aim is to take advantage of the emission characteristics/ properties of the Silicon QDs which can be tuned by their size, as an outcome of quantum confinement (Alivisatos, 1996). The reasons of interest in quantum dots with respect to various other down-converting species such as rare-earth ion complexes (Ledonne et al., 2009, Ledonne et al., 2011) or organic dyes molecules (McIntosh & Richards, 2006; Maruyama et al., 2000; Richards & McIntosh, 2007) are their high stability, brightness and quantum efficiency (Bruchez et al., 1998).

One major benefit of applying a luminescent down-conversion layer onto solar cells is that the luminescence down-converter is merely coupled optically to the solar cell and hypothetically solar cell and spectral down converter can be optimized independently. This concept of down-conversion primarily shifts the photons into the red region from the region (Kanemitsu, 1994; Kanemitsu et al., 1997).

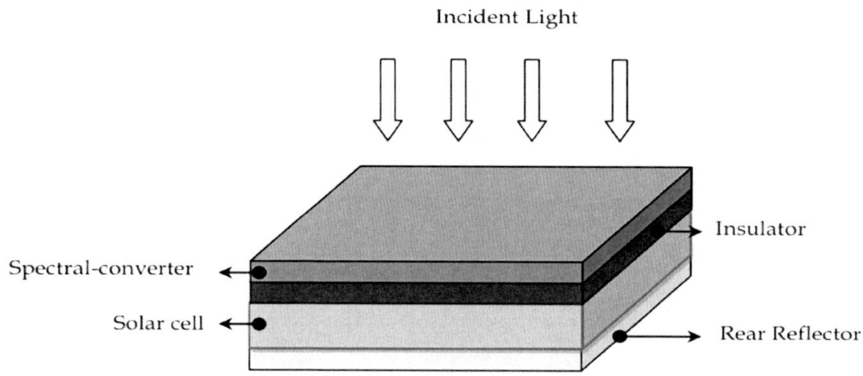

Figure 2.8. Schematic illustration of a photovoltaic down-converter device

[*Source*: https://www.intechopen.com/books/quantum-dots-a-variety-of-new-applications/silicon-quantum-dots-for-photovoltaics].

Many researchers (e.g., Švrček et al., 2004) have observed an increase in efficiency of almost 0.4%. In this scenario, the Silicon QDs were fabricated ex-situ (grinding of electrochemical etched porous silicon) implanted into a SiO_2 based solution and afterward onto a standard silicon solar cell (Švrček et al., 2004; 2010). According to their claim, an improvement in the Internal Quantum Efficiency (IQE) has been observed in the region 300–500 nm (Wei et al., 2010). This difference noted in the performance of solar cell points out the complexity in comparing the efficiency of silicon QDs in various conditions (Sychugov et al., 2005).

Pi et al. (2011) have reported an increase in efficiency of around 0.6% which states the improvement in efficiency observed by Svreck et al. (2004) might have been because of the improved optical coupling of the light in the cell via Silicon QDs implanted in the spin-on-glass at the surface of solar cell (Lehmann & Gösele, 1991; Özdemir & Yakuphanoglu, 2006; Pi et al., 2011). As per the claim of authors, the porous nature of the deposited film could be at the origin of the observed increase in efficiency (porous silicon layers are popular as an anti-reflecting coating (Westland et al., 1988; Chaoui et al., 2008) instead of a down-conversion effect of silicon QDs. Pi et al. (2011) in their paper have proposed to deposit a silicon QDs based ink at the surface of solar cell through spin-coating.

During the recent years, several studies have been carried out on the Si QD based downs-shifters: (notable results have been summarized in the Table 2.1).

Table 2.1. Performance Results of Si QDs Based Luminous Down-Shifters

Solar cell	Quantum dots	Host material	Performance difference [%]	Illuminating spectrum	Reference
c-Si	Silicon QDs	SiO_2	No improvement	AM1.5G	Jestin et al., 2010
a-Si	Ideal QDs	Ideal plastic	No improvement	AM1.5G	Sark et al., 2004
c-Si	Silicon QDs	Spin-on Glass	+0.4	AM1.5	Svrcek et al., 2004
Poly-Si Silicon QDs	Directly on cell	+10	Visible light		Stupca et al., 2007
c-Si	Silicon QDs	Organic solvent	+0.6	–	Pi et al., 2011

The production of silicon QDs is carried out using a novel synthesis approach (Mangolini et al., 2005) and afterward dispersed in an organic solvent. An enhancement of around 2.3% has been observed in short circuit current (Isc) in this case, and therefore constant with the fact that the Silicon QDs film stimulates improvement of light absorption and consequently an increase in the efficiency of the cell. Lastly, Stupca et al. (2007) through his experiments demonstrated the deposition of Si-QDs dispersed in a volatile solvent on an industrialized polycrystalline solar cell. Depending on their size, the Si-QDs which were used showed either orange/red (2.85 nm) or strong blue (1 nm diameter) photoluminescence. A notable improvement of around 70% has been reported, shining blue light on the cells (the IQE of the cell is in this spectral area is poor) and a still noticeable improvement in efficiency of about 10% had been obtained for visible light. Present results on Si-QDs as downshifters permit us to look forward to improvements up to a considerable extent in efficiency by employing this approach within the next years (Meillaud et al., 2006; Song et al., 2008).

2.4.2. Hot Carrier Solar Cells and Generation of Multiple Exciton in Silicon QDs

In this section, two concepts have been addressed which try to avoid the losses which occur due to thermalization of hot carriers. First is the hot carrier solar cells (HCSC), which focuses on the extracting the hot carrier prior to their relaxation to the band edges; second method is based on the concept of generation of multiple excitons (MEG) by using high energy excitons that are observed in a series of QDs (Solar Generation, 2008).

The Figure 2.9 presents a hot carrier solar cell (HCSC) that were proposed in early 80' as a capable approach for increasing the efficiency of traditional solar cells (Ross & Nozik, 1982; Takeda et al., 2009; Würfel et al., 2005). When thermalization of photo-excited is carried out that involves lattice phonons, it significantly reduces the number of carriers that can be separated effectively through the p-n junction, therefore, impeding to attain high voltages in a photovoltaic cell. A prospect of decreasing the thermalization time, which is acknowledged as the major loss mechanism in photovoltaic cells, would permit gathering more carriers when they are still excited to higher energetic states and hence increase the voltage of the cell. Generally, in solar cells based on bulk semiconductor, the damping of the thermalization rate is accomplished at extremely high intensities (order of 1000 suns) (Tsybeskov et al., 1994; Würfel, 1997). Conversely, in low dimensional nanostructures, for instance 1D super-lattices, the larger time of translation with respect to the lattice period of a regular crystal leads to a much smaller Brillouin zone and, resultantly, in extremely small number of acoustic phonon modes, which prevents fast thermalization of excited carriers (Nozik et al., 1990; Westland et al., 1988). As already explained above, material systems of quantum dots and semiconductor nanocrystals have the benefit of improving the efficacy of the absorbing layer of the photovoltaic cell.

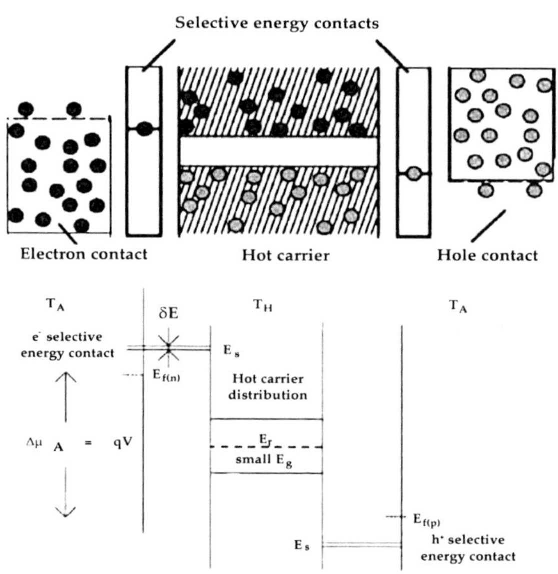

Figure 2.9. The representation of a hot carrier photovoltaic cell with energy selective electrodes (Conibear et al., 2006 [*Source*: https://www.intechopen.com/books/quantum-dots-a-variety-of-new-applications/silicon-quantum-dots-for-photovoltaics].

Talking about HCSC, a quantum dot system can play a significant role also in the realization of energy selective contacts (Conibeer et al., 2006; Conibeer et al., 2008). The crucial concept behind this idea is to increase the efficiency of an HCSC device by gathering the hot carriers over an extremely small energy range with carefully selected energy contacts (SEC). Conibeer and his team (Conibeer et al., 2006) introduced the quantum dot scheme for selected energy contacts where a double-barrier resonant tunneling structure has been described (in this case, a QD-based layer having a discrete energy level is implanted between two insulating layers). An electrode made of such a configuration is supposed to result in strong peaks of conduction at a discrete level of energy (Figure 2.10).

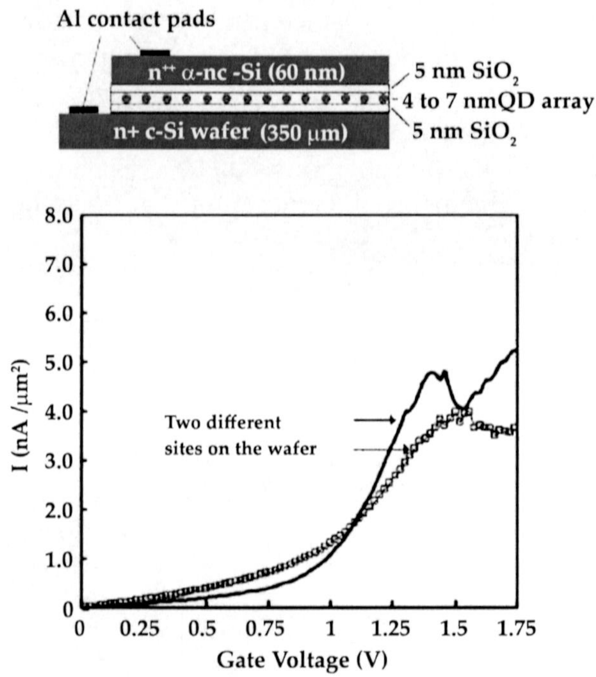

Figure 2.10. The sketch of an SEC experiment (on top) and I-V curves at room temperature displaying the NDR features (Conibeer et al., 2008). [*Source*: https://www.intechopen.com/books/quantum-dots-a-variety-of-new-applica-tions/silicon-quantum-dots-for-photovoltaics].

Proof-of-concept experimental investigations have been carried out by using nanocrystals of silicon embedded as double-barrier electrodes in a silica matrix (Conibeer et al., 2008). The double barrier resonant tunneling architecture used by Conibeer et al. (2008) comprised of a 4 nm thick layer of Si-nanocrystal sandwiched between 2–5 nm barrier layers of SiO_2. Si-nanocrystals have been developed through an elevated temperature treatment of the multilayer architecture, which is known to instigate a phase separation of SiO_2 and Si inside the silicon abundant silica layer. The thickness of layer controls the sizes of the Si-nanocrystals which in its turn is offering the specific distinct energy levels inside the double tunneling barrier layer. The noteworthy outcome of this study is the depiction of negative differential resistance (NDR) at room temperature in the I-V curve of the electrode. In spite of the poor NDR resonance witnessed in the Si nanocrystals experiments, the results for 1D energy selection are extremely encouraging. Recently, modern SI based solar cells with Si nanocrystals-based energy selective electrodes are being explored in much detail to address the structural, spectroscopic and compositional characteristics of the nanocrystalline systems (Aliberti et al., 2010; Shrestha et al., 2010).

The concepts of employing Si nanocrystals-based energy selective electrodes have offered a critical potential to enhance the efficiency of silicon solar photovoltaic cells. However, further research on considerably improved NDR resonance quality and properties is expected to excel shortly.

CHAPTER 3

FABRICATION PROTOCOLS FOR QUANTUM DOT SOLAR CELLS

CONTENTS

3.1. INTRODUCTION

Colloidally produced quantum-confined semiconductor spherical nanocrystals, often known as quantum dots (QDs), presents a high degree of electronic, chemical and optical tunability. As a consequence, there is a growing interest in using colloidal QDs for optical and electronic optical applications which are reflected in abundant publications in the field of quantum dots. This chapter particularly addresses the details regarding the techniques and processes for synthesis QD based photovoltaic cells specifically using PbS and PbSe QDs (Brown et al., 2014; Chuang et al., 2014; Ning et al., 2014). The details concerning the QDs are a depiction of knowledge acquired through experimental investigations employed for fabrication of QD photovoltaic cells. Successful fabrication of solar cells requires experience and deep knowledge regarding the source materials. This chapter is intended to provide information on the acceleration of that fabrication processes (Goodwin et al., 2014; 2016; Marshal et al., 2016). The protocol addressed in this chapter is intended to be a starting point for fabricating PbSe and PbS QD test bed photovoltaic cells. Various steps for development of conductive QD films using dip coating and spin coating are also discussed in detail. Finally, the discussion of protocols which involve the synthesis of PbSe and PbS QDs through cation exchange reaction for high-efficiency solar cells is also included in this chapter (Klimov et al., 2016; Li et al., 2016; Xu et al., 2016).

Colloidal semiconducting quantum dots offer a unique platform for manipulating a wide variety of optoelectronic devices due to their quantum confinement and improved surface-to-volume ratios. These distinctive size-dependent characteristics allow for emergent optical and electrical phenomenon resulting from QD–ligand and QD–QD electronic coupling. Benefits of QDs include excellent optical transitions (i.e., light emission and absorption), excellent control over work-function and band edge energies, a wide range of tunable bandgap energies, facile fabrication, and facile incorporation into matrices or development as thin films (Schaller & Klimov, 2004; Ellingson et al., 2005; Murphy et al., 2006). Moreover, QDs also offer desirable excited state characteristics which include multiple exciton generations (MEG), long carrier lifetimes along with energy and charge transfer characteristics with surface adsorbates (De Geyter et al., 2010; Lee et al., 2010; Pak et al., 2012). Thus, QDs are being investigated and commercialized as dynamic components in displays (televisions, computer monitors, etc.), bioimaging, optical detector applications, and green energy applications which include photovoltaics, solid-state lighting and thermoelectrics (Padilha et al., 2013; Cirloganu et al., 2014; Weidman et al., 2015).

Photovoltaic cells synthesized from quantum dots have the potential to offer higher power conversion efficiencies because of enhanced MEG (Chen et al., 2013; Milliron, 2014; Khan & Bowden, 2016). Although MEG is augmented in typical quasi-spherical quantum dots over bulk substances and traditional thin film semiconducting materials (Konstantatos et al., 2012; Shirasaki et al., 2013; Supran et al., 2013). The threshold and efficiencies of the MEG phenomenon can be improved via shape control, internal quantum dot heterojunction interfaces, and additional material investigation (Ballou et al., 2004; Muller et al., 2010; Sun et al., 2012). Research toward enhancing the MEG efficiency is trending nowadays. However, to harness the benefit of these unique features of quantum dots, researchers also need to manufacture prototype energy conversion structures which can serve as a testbed for innovative nanoscale phenomena (Gao et al., 2004; Kim et al., 2004; Lan et al., 2016). In photovoltaic cells, the active layer usually utilizes electronically coupled quantum dot arrays. However, QDs sensitized structures have exhibited significant potential as well (Parker & Singh, 2010; Wang et al., 2012; Liu et al., 2016). The ultimate performance of a solar cell depends upon a range of correlated factors which are listed below:

- The QD synthesis.
- Degree of QD–ligand and QD–QD electronic coupling.
- QD film morphology.
- QD surface passivation and several carrier recombination centers.
- Appropriate contact layers.

The current chapter contains protocols which provide detailed procedures for synthesis of standard QD photovoltaic cells using PbSe and PbS which can approach the power conversion efficiency of up to 8–9% (Zhang et al., 2012; Moroz et al., 2014; Supran et al., 2015).

3.2. QUANTUM DOT SOLAR CELL PROTOCOL

Some major raw materials and fabrication steps in quantum dot photovoltaic cell protocol are addressed in the following text.

3.2.1. Reagents

MPA (3-Mercaptopropionic acid, >99%), DMF (N,N-dimethylformamide, anhydrous, 99.8%), 1,2-ethanedithiol (EDT, ≥98%), methanol (reagent grade), PbI2 (Lead(II) iodide, 99.9985%) titanium(IV) ethoxide (≥97%), acetonitrile (99.8%), MoO3 (molybdenum trioxide, 99.99%), and ACS reagent (hydrochloric acid, 37%) are typically used as raw materials for synthesis of PbSe and PbS based quantum dots (Kwak et al., 2012; Mashford et al., 2013).

3.2.2. Substrates

Polished float glass substrates (dimensions: 25 mm × 25 mm × 1.1 mm) with pre-patterned ITO or FTO on 1 side are utilized for the manufacturing of Thin Film Devices (Dai et al., 2014; Gong et al., 2016). Synthesis of Ligand Exchange Solutions

The ligand exchange solutions utilized for the synthesis of the QD absorber layer incorporates 10% MPA of volume in methanol, 1 mM EDT in acetonitrile and ten mM PbI_2 in DMF. The stock of PbI_2 powder is stored in a glovebox, and only the required quantity is removed. The mixing of PbI_2 in DMF is carried out in the air to develop the ligand-exchange solution; however, the dissolution can be carried out inside the glovebox (Hanna & Nozik, 2006; Smith & Michl, 2010; 2013). The blend is sonicated for approximately 10−30 min to dissolve the solid completely. Variations in the properties of PbI2 are observed for PbI2 products obtained from different vendors, and certain lots of PbI2 do not easily dissolve in the DMF. Hence, it is advisable to use high purity PbI2 for the synthesis of Pb based quantum dots. For the synthesis of the 1 mM EDT solution, the glovebox is used for the preparation of 1 M EDT in acetonitrile solution followed by its removal in the air for achieving the desired dilution for 1 mM solutions. The >99% MPA solution is exposed to air, and the diluted 10% if volume in methanol solution is synthesized in the air (Semonin et al., 2011; Ip et al., 2012; Böhm et al., 2015).

3.2.3. Solar Cell Fabrication

The entire process of solar cell manufacturing (Figure 3.1a) starts with a glass substrate which is pre-patterned with ITO or FTO (Figure 3.1b: slightly gray areas are FTO). Our pattern is comprised of a rectangular line (racetrack) in the center, and 6 pads are present on the sides. Other patterns can be developed using the general protocol established here (Bartnik et al., 2010; Cunningham et al., 2011; Davis et al., 2015).

Metal bus bars are used as a coating on the FTO top for minimizing the resistivity to extract the current in the affected FTO areas, and in between the 6 central areas which will become 6 individual pixels of a solar cell, as shown in Figure 3.c. Incorporation of metal bus bars results in increased fill factor, by decreasing the series resistance of the employed substrate. After evaporation of bus bar, the deposition of TiO_2 takes place (Figure 3.1d), followed by quantum dot layer deposition. Zinc oxide (ZnO) can be employed in place of TiO_2 as an n-type layer and electron transport layer. However, it has been observed that ZnO based solar cells offer significant irreproducibility in performance; therefore, TiO_2 is extensively employed for this reason (Price et al., 2015; Du et al., 2016; Yang et al., 2016). The high annealing temperature of TiO2 (i.e., 450°C) suggests the predominant use of FTO due to its favorable temperature stability in contrast to ITO. ITO exhibits an increase in its resistivity with annealing at temperatures higher than 200°C, which results in enhanced series resistance of the solar cell.

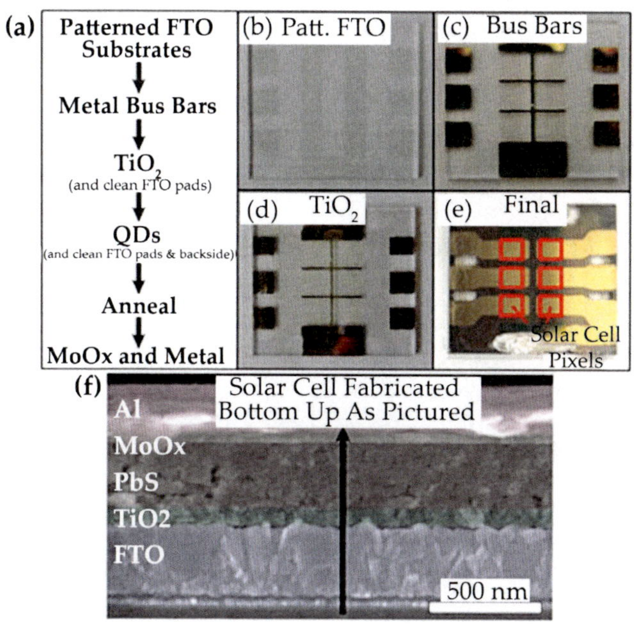

Figure 3.1. (a) Schematic illustration of flow diagram containing the steps involved in the development of a QD photovoltaic cell. (b) FTO patterned Glass substrate present as a rectangle in the center, which is called the "racetrack,"

and 6 pads are present on 2 sides. (c) The substrate after bus bar deposition of metal (Gold in the photograph). (d) The substrate after deposition of TiO2. The lighter patches at, the lower and upper pad regions, as shown in the image, exhibit the pattern of the TiO2 removal using ethanol and the cotton swab. (e) A finished photovoltaic cell with "finger" Au contacts. Each of the 6 red boxes demonstrates an active area for a distinct pixel of the solar cell, which is distinguished by the overlap of the bottom FTO and top metal contacts. The removal of QD active layer has been carried out from the large metal/FTO pads in the lower and upper portions of the substrates as shown in the image. Finally, indium solder was incorporated into the lower and upper portions of the metal/FTO pad regions for a reduction in contact resistance. (f) False color scanning electron microscope image is displaying the cross-section of a completely developed solar cell. [*Source*: https://pubs.acs.org/doi/abs/10.1021/acs.chemmater.6b02939].

Deposited QDs and TiO_2 are chipped-off the lower and upper pad regions of the racetrack (FTO) to function as the bottom electrical contacts (i.e., soldered regions in lower and upper portions of the substrate in Figure 3.1e).

Lastly, metal (Au or Al) and MoO_x are deposited by thermal evaporator using a mask to form 6 contact lines called fingers (Figure 3.1e); thereby; forming 6 individual solar cell pixels (0.101 cm^2). Electrical contacts are made to on top of every solar cell pixels by contacting the corresponding metal/FTO pad on the right or left side of the substrate, as oriented in Figure 3.1e (Zhao et al., 2015; Zhang et al., 2014; 2015).

3.2.4. Substrate Preparation and Deposition of TiO2 Layer

- The substrates are vigorously cleaned with an ethanol-soaked tissue. The tissue is typically sprayed with ethanol and then dried instantly with a pressurized stream of N_2 or air.
- Gold or silver bus bars (with 75−100 nm thickness) are deposited by thermal evaporation technique using a shadow mask. The evaporation rate is not critical; typical deposition rate is 0.5−2 Å/second. Bus bars must be smaller than 100 nm to restrict their effect on the morphology of TiO_2 film.
- The sol−gel precursor of TiO_2 is synthesized by mixing 125 μL of deionized water, 2 drops of hydrochloric acid (HCL) and 5 mL of ethanol in a beaker. Approximately 375 μL of titanium (IV) ethoxide (stock) is added dropwise. Stock titanium ethoxide

solution is typically stockpiled in a glovebox, and only the required amount is released to the air. The resultant product is a clear liquid without any particulate species. The solution is then covered and stirred for about 1 hour before storing in a freezer. The sol–gel is taken out of the freezer 10 minutes before use and stirred until it achieves equilibrium at room temperature (Hu et al., 2004; Wu et al., 2014; Lee et al., 2015). The mixture of sol–gel should be discarded if it contains visible particulates or if the solution is noticeably yellow cloudy.

- The synthesis of the TiO_2 layer is carried out by spin coating technique using 70 μL of the sol–gel mixture for at 1400 rpm for 30 seconds. The sol–gel layer of TiO_2 is removed from the lower and upper metal/FTO contact pads, and the smaller 3 right and 3 left metal/FTO pads, as oriented in Figure 3.1d. A cotton swab soaked with ethanol is used for removal of TiO_2 instantly after the completion of the spin cycle. The cleaning of FTO/metal pad must be carried out immediately after spinning process, and before drying of the sol–gel solution. Removal of the sol–gel layer of TiO_2 is critical over the large lower and upper metal/FTO pad areas. The other 6, smaller, metal/FTO pads on the sides are coated with the metallic top contacts to facilitate the cleaning of small FTO. After wiping the metal/FTO pads, each substrate is placed on a hot plate at around 115°C, to remove water particles, while the next substrates re prepared. After the finishing of the last substrate, the substrates are placed on the hot plate for further 20 min (Vinodh et al., 2003; Church et al., 2013; Wheeler et al., 2016).
- Finally, the substrates are positioned in a preheated furnace for 30 min at 450°C, regardless of the substrate nature, i.e., ITO or FTO coated the glass. The coated films are then stored in open air (without light) before deposition of a QD film (Turyanska et al., 2009; Gao et al., 2013).

3.2.5. Deposition of QD Layer

The quantum dot layer is typically deposited using multiple dip coating or spin coating cycles. The number of dip or spin cycles along with the concentration of quantum dots determines the thickness of QD film. Both approaches are highly tunable (i.e., controlling ligand exchange and layer thickness patterns) and both result in photovoltaic cells with comparable performance. The choice of the method normally depends on the availability of the equipment (Osterwald, 2003; Emery, 2011; Snaith, 2012). Dip coating needs enough total volume to facilitate the immersion of substrate. Therefore, the concentration of QD solution is typically kept lower in case of dip coating (i.e., 10–20 mg/mL), as compared to the concentration in case of spin coating (i.e., ~30 mg/mL). Consequently, dip coating is commonly a slower deposition technique, because each coating cycle deposits fewer number of quantum dots, but this method is usually desirable for achieving fine control of thin QD films and for the potential to fabricate more photovoltaic cells from a single cycle of QD synthesis (Figure 3.2).

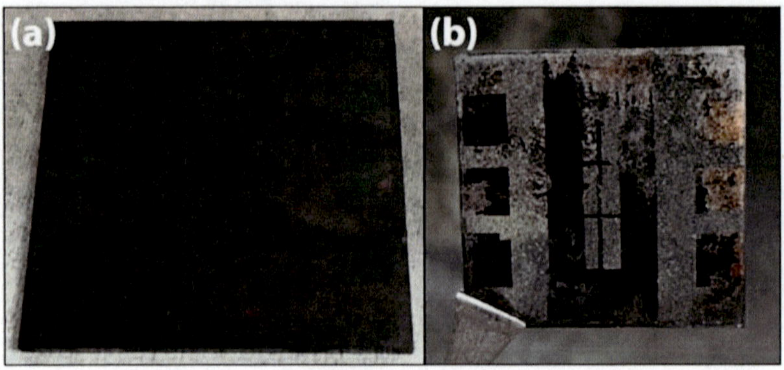

Figure 3.2. (a) Typical film obtained after spin coating. (b) Film showing peeling effect during the MPA treatment [*Source*: https://pubs.acs.org/doi/abs/10.1021/acs.chemmater.6b02939].

Higher solution concentrations will result in deposition of thicker layers per cycle followed by a quick synthesis of solar cell structure (Johnston et al., 2008; Law et al., 2008). On the other hand, lower concentrations yield thin layers followed by slow solar cell manufacturing. A drawback of the spin coating process is that some of the sol-gel solutions are wasted as it is spun-off the FTO/ITO substrate. The use of tweezers is highly recommended for the handling of prepared solar cell structures (Sites et al., 1990; Gloeckler & Sites, 2004).

3.2.6. Spin Coating Protocol

Before spin coating, the quantum dots are distributed in octane at a concentration of around 30 mg/mL. The solution is usually filtered through a 0.2 μm PTFE or Nylon filter to eradicate agglomerates and enhance the film uniformity (Nadenau et al., 2000; Hegedus et al., 2004; 2007).

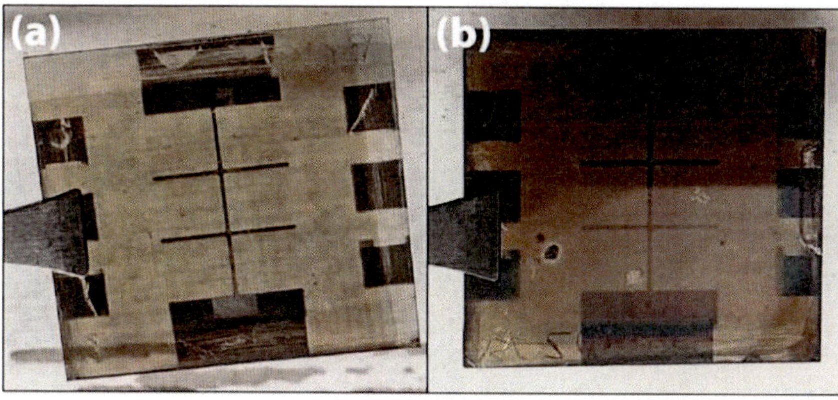

Figure 3.3. Spin-coated films (a) The film after ~25 nm thickness (b) ~ 150 nm thick EDT-treated PbSe quantum dots. [*Source*: *https://pubs.acs.org/doi/abs/10.1021/acs.chemmater.6b02939*].

The glass side of the substrates is cleaned in every deposition scheme. Imperfections may arise during the coating process. For instance, a dark spot is present in the bottom-left (in Figure 3.3b) near the tweezers that arises as a result of QD-agglomerates which float in the ligand exchange solution. This dark spot is not present in an active photovoltaic area, however, but it can develop in any active region as well. However, some defects may not have a substantial impact on the performance of a pixel, such as the light spot over the middle-right pixel of the Figure 3.3b (Li et al., 2011; Thomson et al., 2011; Nan et al., 2012).

The steps involved in a spin coating procedure, for deposition of films, are illustrated in the following text.

- 80 μL of quantum dot solution, enough to shield the central region of the substrate encompassing 6 active solar cell regions, is dispensed onto the substrate using a micro-pipette. The solution is spread uniformly to cover the whole surface. If the air bubbles form, it is essential to remove the bubbles by perturbation of the solution via air puffs using the empty micro-pipette (Pu et al., 2013; Yang et al., 2015).
- The spinning of the substrate is performed at 2000 rpm for 30 seconds, with a 3-second ramp, for obtaining a glassy film along with dried octane. The excess liquid present on the substrate back

must be dried using a scientific tissue before the ligand treatment.

- The quantum dot film is immersed in the ten mM solution of PbI_2 for 3 minutes to facilitate complete exchange.
- After removal of the substrate from the DMF solution, tapping of the film is performed against the beaker-side to allow the excess DMF diffuse back into the available ligand solution.
- Excess PbI_2 and DMF are removed from the film by submerging it several times in a neat acetonitrile solvent. It is imperative to adjust the film with its edge in the acetonitrile followed by the opening of the tweezers to discharge the DMF trapped between the substrate and the tweezers.
- A gentle stream of air is used to dry the film to assist the removal of acetonitrile from the film edge. The above 6 steps are performed 4 times to construct the PbI_2 layer (Figure 3.2 a).
- Spin coating of the MPA (3-Mercaptopropionic acid, >99%) treated layer is carried out in the same fashion as illustrated above, except that MPA with 10 volume % in methanol is utilized as the ligand-exchange solution.

The spun film is gradually lowered into the MPA solution and held for ~1 second followed by steady removed and rinsing of excess MPA in neat methanol. The drying of the film takes place using a gentle air-stream (exposure time of ~4 seconds). This process is repeated twice to develop the MPA layer with a desired thickness. Inadequate rinsing off the excess MPA solution may result in the development of cloudy white regions upon deposition of the next QD layer from octane (Luther et al., 2008; Ma et al., 2009; Pattantyus-Abraham et al., 2010). This effect can severely impact the pixels rendering them Ohmic (unable to exhibit rectifying performance). It is, therefore, recommended to discard the film if it contains too many affected pixels. The film should be rinsed several times with neat methanol to avoid such problems. Decreasing the concentration of the MPA up to 1% of volume in methanol can rectify the problem of peeling which arises during the of MPA treatment (Figure 3.2 b). Peeling of the film during the MPA treatment is severe if ZnO (rather than TiO_2) is used; therefore, MPA with lower concentration is recommended to cater the problem. It has been established that the variations in the MPA concentration between 1 volume percent to 10% in volume do not result in substantial differences in the performance of a solar cell (NREL, 2015; Jeffries et al., 2016; Wang et al., 2017).

An alternate spin coating methodology involves the direct execution of the ligand treatments on top of the film while keeping the substrate in the spin coating equipment. This technique differs from conventional spin coating approach because of the longer soak times in the DMF/PbI_2 solution. The modified technique allows the scientists to process 4 thin films simultaneously by permitting the immersion of 3 films in ligand solution while the QD layer gets deposited on the fourth thin film. A four-slot substrate holder is used for clamping the substrate which fits inside the beaker for immersing substrates in the DMF/PbI_2 solution. Although the quantum dot layers can be deposition directly on the film using the spin coater, the above-described soaking route provides excellent results.

3.2.7. Dip Coating Protocol

In dip coating process, the quantum dots are distributed in hexane at a certain concentration (~15 mg/mL). The quantum dots are filtered using a 0.2 μm PTFE or Nylon filter to remove agglomerates before the deposition process. Ligand solutions which are reused from previous depositions need to be filtered as well. The empty beakers are covered with a watch glass (during deposition) to slow the evaporation phenomenon and block dust particles. A table is made on the window of the fume hood to indicate the dip counts via tick marks for keeping track of film thicknesses. Traditionally, quantum dot solar cells are dip coated using EDT ligand treatment. Dip coating of QD layers with EDT or similar organic ligands typically follows the appending guidelines (Choi et al., 2009; Leschkieset al., 2009; Backhouse et al., 2011).

- The substrate is gently lowered in a vertical position into the quantum dot solution followed by its gentle removal from the solution. The lowering of the substrate in the solution must be done carefully to avoid the wetting of the tweezers. The diffusion of the solution in the tweezers may cause a droplet to flow down the freshly formed film and upset uniformity upon the removal of the substrate. The deposition of QD layer takes place during the withdrawal of the substrate, so the lowering-speed is not critical. A decent withdrawal speed is roughly 2 seconds to remove a substrate of 1 in. length (~1.3 cm/s). Quick removal of the substrate results in thicker films due to trapping of more solution on the substrate; however, homogeneity and uniformity of the film are typically low. The non-uniform film can be treated by immersing it again into the quantum dot solution. It is essential to carry out this step before ligand exchange phenomenon. The substrate must be kept immersed in the QD solution for around ~5−15 seconds to allow for uniform dispersion on the non-uniform layer (Kramer et al., 2010; Pattantyus-Abraham et al., 2010: Gao et al., 2011).

- When the solution is removed from the substrate, the bottom edge of the substrate is tapped onto a paper towel to remove any excess quantum dot solution. If permitted to dry, the excess quantum dots may flake-off during the phenomenon of ligand exchange. The hexane must be fully dry before dipping of the substrate into the ligand solution.
- The dipping of QD-coated-substrate into the ligand solution (1 mM EDT in acetonitrile) takes place after hexane drying. Comparable dipping speeds are employed for the ligand exchange process, but a 1-second soak is included before removal while the film is immersed.
- The film is allowed to dry after its removal from the EDT solution. Compressed nitrogen gas is used to blow-off the droplets of solution entrained on the film.
- After drying the film, it is rotated 90° (and so on) and then dipped again into the QD solution. Turning the substrate at 90° account for less directional non-uniformities.
- It is recommended to use a profilometer for calibration of the QD film thickness per dip cycle. Profilometer facilitates the researcher to determine the extent of film deposition up to a certain thickness.

Dip coating is also used to synthesize QD solar cells using the amalgamation of MPA and PbI_2 ligands treatments demonstrated in the spin coating section; however, the process is slightly altered:

- Dipping of the substrate into the DMF/PbI_2 solution takes place at the rates similar to those employed in the case of EDT. However, a soak time of 10 seconds is incorporated before the removal of the substrate from the ligand exchange solution.
- Immediately after the substrate removal from the PbI_2 solution, excess PbI_2 are DMF removed from the substrate by dipping it into neat acetonitrile. In the submerged condition, the tweezers are intermittently opened and closed to remove excess DMF/PbI_2 from the tweezers as well.

- The substrate is then dried using the compressed clean air of compressed nitrogen immediately after its removal from the acetonitrile rinse. It is also advisable to lay down the substrate and remove the trapped-liquid from the tweezers after clamping them onto a tissue.
- The substrate is then rotated at 90°, and the procedure is repeated.
- The quantum dots do not adhere perfectly to the glass-side of the substrate after the ligand exchange process. The quantum dot film on the backside can be disrupted and flaked into the DMF/PbI_2 or acetonitrile solution. The flaked quantum dots float on the solution top. An oily residue comprising of the oleate ligands may also form in the solutions. The oils and flakes may be removed by cautious scooping using a folded scientific tissue. On the other hand, the acetonitrile rinse or PbI_2 solution may be substituted. The flakes must be minimized to avoid their entrapment in the QD film. Usually, the QDs show a reduced flaking rate as the thickness of the film increases on the substrate-back. Reduction in the PbI_2 concentration (i.e., 1 mM rather than ten mM) will decrease the film flaking rate. If the layer thickness of QDs per dip is less than 20 nm, the use of DMF/PbI_2 concentration in the range of 1 to 10 mM is suitable without increasing the soak time. The reduction in the PbI_2 concentration is suggested if the QDs exhibit is flaking off the substrate. The rate of flaking may decline with the increase in film thickness, but inhomogeneity may appear by the initial flakiness (Beard et al., 2010; Mohan et al., 2016).
- Before the deposition of QDs with the second ligand (i.e., MPA), calculation of the average thickness-per-dip is made using a profilometer to calculate the required number of dips for achieving the desired thickness. This is mostly done to maintain the constant thickness of the absorber layer when the aim is to compare the performance of solar cells produced from different QD batches.
- After the finishing of PbI_2-treated layers, the MPA ligand process is performed in the way similar to the process described in step 7 of the typical spin coating procedure. However, the deposition of each fresh layer of QDs is carried out by dip coating instead of spin coating.

In the case of photovoltaic cells containing quantum dots with MPA ligands and PbI_2, the thickness ratio of the entire PbI_2-treated layer and the whole MPA-treated layer is typically 2:1 (i.e., 200 nm of PbI_2-treated quantum dots and 100 nm of MPA-treated quantum dots).

After deposition of the QD film, the glass side of the substrates is cleaned with cotton swabs soaked in certain solvents, e.g., methanol, acetonitrile or ethanol. On the active side, the large lower and upper metal/FTO contact pad areas must be cleaned cautiously such that the solvent or cotton does not touch the area of 6 pixels. The six-side finger pads are also cleaned carefully (Chuang et al., 2014; Yuan et al., 2016).

After cleaning of the FTO pads, annealing of the films takes place in a glovebox filled with N_2 (H_2O < 1 ppm, O_2 at <5 ppm) at 110° C for 20 minutes. Annealing optimization can be performed for different QD-ligand combinations (Oh et al., 2013; 2014).

3.2.8. Top Contact Layer and Completion of the Solar Cell

After annealing, the deposition of the back-contact layers on the film is carried by a thermal evaporation technique. The deposition of 100–150 nm of Au or Al and 15–20 nm of MoO_x is required for the back contacts. For deposition of films, the metallic or MoO_x powders (pellets) are loaded into a resistive heater, and the chamber is vacuumed to a base pressure of 10^{-7} – 10^{-8} Torr. The MoOx is evaporated gradually at a typical rate of 0.02 nm/s. The Au or Al is then deposited at a rate of about 1.5 nm/s. These depositions are performed in a thermal evaporator having multiple sources to avoid the exposure of the film to the atmosphere between Al or Au and MoO_x depositions (Crisp et al., 2015; Luther, 2015).

A false-color cross-sectional SEM (scanning electron microscopy) image of a finished solar cell (Figure 3.1f) illustrates highly smooth and uniform QD films which are a depiction of the above-described protocols' execution. The completed solar cells are kept in a drawer under controlled conditions.

3.3. CHARACTERIZATION OF A SOLAR CELL

There are different characterization techniques for investigating the properties of the solar cell. Some of the techniques are discussed below:

3.3.1. Current−Voltage Response

Current density vs. voltage (J−V) characterization of the photovoltaic cells is carried out in an oxygen-free atmosphere using an Oriel Sol3A (94043A) solar simulator system equipped with an AM1.5G filter and a 450 W xenon arc lamp. Measuring the J-V of solar cells in air results in the reduction of fill factor, which improves when the cell is reverted to an inert atmosphere. A metal plate is utilized to aperture each of the 6 pixels throughout the characterization process to prevent illumination of the areas outside the defined active regions, which results in artificial inflation of the PCE (27–28). It can be noticed that the J−V measurement of any single pixel can unveil the illumination of the other 5 pixels held at the open circuit. The total aperture area (i.e., 0.101 cm^2) or the pixel area is used to compute the current density under illumination and in the dark, respectively. The height of the photovoltaic cell stage is tuned until the illumination intensity reaches 100 mW/cm^2, as calculated by the current output on a calibrated silicon reference photodiode. A mismatch factor can be applied to correct the value of short circuit current (J_{sc}) as appropriate contingent to the cell responsively and QD size. Figure 3.4 illustrates an exemplary J−V curve for photovoltaic cells with PbS (red) or PbSe (blue) QDs. Although QD solar cells do not display the hysteretic behavior witnessed in perovskite solar cells, stress effects (developed during measurement) have been observed in QD solar cells, e.g., reduction in V_{OC} due to illumination-soaking under open circuit conditions. An average standard deviation of ±10% is observed for the efficiency of a solar cell. Deviations in efficiency values of solar cells arise due to measurement stress effects and inhomogeneity across the substrate area (Ko et al., 2014; Marshal et al., 2015; 2016).

3.3.2. Characterization: Spectral Response/Quantum Efficiency

The external quantum efficiency or EQE measurement is employed for characterization of current losses in a photovoltaic solar cell concerning wavelength or photon energy. EQE is the measure of the electric flux, typically measured as current density of a photovoltaic cell concerning the incident flux (photonic flux). The EQE is also known as the incident-photon-to-current-efficiency or IPCE (Figure 3.4).

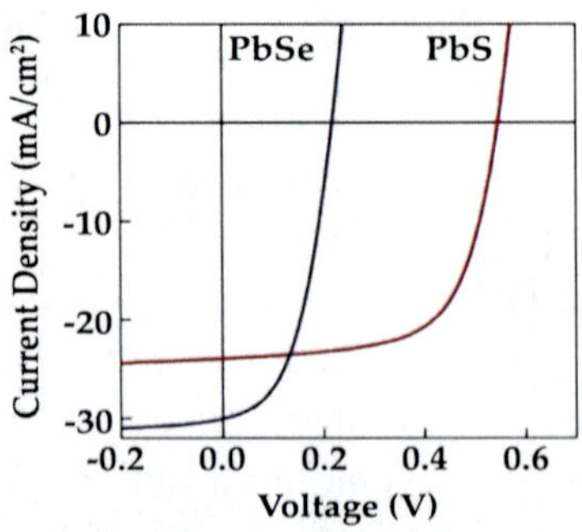

Figure 3.4. Example of samples depicting J–V response of PbS (1.2 eV) and PbSe (0.78 eV) [*Source*: https://pubs.acs.org/doi/abs/10.1021/acs.chemmater.6b02939].

The fill factor, efficiency Jsc, and Voc of the PbS solar cell are 64.0%, 8.36%, 23.9 mA/cm2, and 0.546 V, respectively. The fill factor, efficiency Jsc, and Voc of the PbSe solar cell are 48.2%, 3.15%, 30.0 mA/cm2, and 0.217 V respectively. The structural design of these photovoltaic cells was as displayed in Figure 2.1.

EQE losses typically include IR absorbance and parasitic UV in the electron or hole-extracting layers (e.g., TiO_2, MoO_x or FTO), interface reflections, insufficient absorber thickness (resulting in IR loss), and diffusion lengths of small carriers in the quantum dot layer or electron/hole-extracting layer. The measurement equipment contains a light source, light filters, a light guide, monochromator, a chopper, a lock-in amplifier, probe setup for current measurement, and a preamplifier. More information on the measurement, calculation, and instrumentation can be found in the literature (27–31). Another key entity for spectral response measurement is known as internal quantum efficiency (or IQE). One of the solar losses in photovoltaic cells occurs due to reflections, which are taken into account in the IQE measurements. Reflections usually stop some fractions of light from approaching the quantum dots; thereby, resulting in loss of the incident photons without their conversion to current. The IQE can be measured in two distinct ways, i.e., absorber material's IQE or the IQE of a photovoltaic device. In the case of a device, the IQE_{device} is typically equivalent to QE after consideration of the reflective losses in the whole device. Hence, the IQE_{device} is usually measured as the ratio of electron flux to the photonic flux absorbed by the whole dielectric stack which constitutes the solar cell. For instance, Figure 3.5 displays an IQE_{device} and EQE for a PbSe QD photovoltaic cell (Ersley et al., 2012; Gao & Johnson, 2012; Hughes et al., 2012).

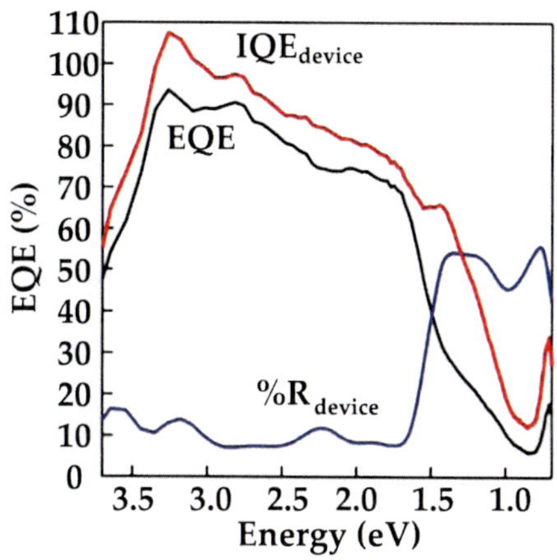

Figure 3.5. Figure 1.5. Example of IQE device spectra and EQE (reflectance) of PbSe QD photovoltaic cells (0.72 eV). An IQE device with a value more than 100% indicates the contribution of MEG towards total collected current (16). The structure of this solar cell is shown in Figure 3.1. [*Source*: https://pubs.acs.org/doi/abs/10.1021/acs.chemmater.6b02939]

The red peak in the above graphs indicates the example a solar cell with greater than 100% IQE_{device} in the UV region due to the contribution of MEG towards the performance of the solar cell (16). In the second case, the $IQE_{absorber}$ becomes QE after taking into account the absorptive and reflective losses in all the solar cell materials except for the quantum dot absorber layer. Conversely, $IQE_{absorber}$ is the measure of the ratio between electric flux and the photonic flux absorbed by the QDs (active layer). Hence, $IQE_{absorber}$ is usually greater than IQE_{device}.

Some characterization techniques permit the measurement of reflectance spectrum (diffuse ± specular reflectance) simultaneously using EQE measurement technique. On the other hand, the reflectance spectra can be attained using UV–visible-NIR spectrophotometers fitted with an integrating ball. UV-visible-NIR is performed by making a QD photovoltaic cell on an unpatterned substrate a way similar to the patterned photovoltaic cells. The deposition of back contacts is carried out onto the entire area of the twin substrate. The reflectance spectra of the photovoltaic device are obtained by thoroughly measuring the twin film reflectance in a UV–visible-NIR spectrophotometer integrated with an integrating sphere (having an incidence angle of 8°) and a specular reflectance using standard NIST calibrated spectrometer. Eventually, the IQE_{device} is measured as EQE/ $(1 - R_{Device})$ (here R_{device} indicates the twin cell reflectance spectrum). Measurement of the $IQE_{absorber}$ necessitates the calculation of the reflectance at all interfaces and the measure of absorbance in all material (Amassian et al., 2011; Tang et al., 2011).

3.4. QDS SYNTHESIS PROTOCOL

All the raw materials, for the synthesis of PbS, PbSe, CdS and CdSe QDs, are typically purchased from Sigma-Aldrich or any other renowned brand and are employed as received. Typical reagents include technical grade Oleic acid (OA), technical grade 1-Octadecene (ODE), $PbCl_2$ (99.999%), technical grade oleylamine (OLA), CdO (≥99.99%), selenium (powder mesh, 99.99%), $(NH_4)_2S$ (40–48 weight % in H_2O), reagent grade hexane, trioctylphosphine (TOP), and reagent grade ethanol (200 proof).

3.4.1. General Synthetic Notes

i. The synthesis of CdS quantum dots is derived from the process suggested by Zhang et al. (2012) but without drying of the sulfur precursor. The synthesis of CdSe quantum dots is modified from that suggested by Nan et al. (2012) and Zhang et al. (2012). The syntheses of PbSe and PbS quantum dots by cationic exchange reactions from CdSe and CdS, respectively, are tailored from the process suggested by Zhang et al. (2012).
ii. At the start of a synthesis process, reagents are degassed before heating. This is done by removing the gases from the flask atmosphere (for 1–2 minutes) using the Schlenk line, or until bubbling has stopped. Gas removal is followed by 3 cycles of N_2 flushing before the heating mantle is switched on.
iii. The base vacuum pressure is kept around 20–100 mTorr. When a reaction is initiated using an injection, 20 mL syringes and 16 gauge needles are used irrespective of the liquid amount being injected. This guarantees a rapid injection, which will enhance the size distribution of the final product.

3.4.2. Synthesis of CdS Quantum Dots

i. The precursor solution is synthesized with 4 g of OA, 18 g of the ODE, and 0.66 g of CdO in a round-bottom 100 mL flask.
ii. The reagents are continuously stirred and heated to 260°C. The degassing of the solution takes place by alternating between N_2 and vacuum while the solution temperature is kept below 80°C. Above

the temperature of 80°C, the flask atmosphere is switched to N_2. To minimize the leakage of air into the flask atmosphere, the nitrogen pressure in the Schlenk line is maintained high enough to facilitate the escape of excess N_2 through the bubbler (a component of the Schlenk line.)

iii. On the contrary, 180 μL of $(NH_4)_2S$ is incorporated to 15 mL of OLA followed by vigorous stirring. This outgassing of mixture takes place aggressively in the absence of capping. The ammonium sulfide forms a gel when supplemented into the OLA and must be allowed to dissolve before injection into the Cd–oleate solution fully. Capping of the vial may be applied temporarily while shaking the solution to facilitate dissolution; however, the cap must be loosened completely afterward.

iv. The reaction is kept undisturbed at 260°C until all of the color (burgundy) has disappeared followed by the clear appearance of solution which indicates the conversion of CdO to Cd–oleate.

v. t is essential for the entire CdO to react before the succeeding step. Gentle shaking of the flask helps to eliminate unconverted CdO from the walls of the flask.

vi. The flask is separated from the heating mantle and permitted to cool up to the temperature of 32°C.

vii. The replacement of heating mantle takes place, and the temperature control is put at 32°C.

viii. The OLA/ammonium sulfide solution is then added to the flask for reaction in 1 hour. The reaction yield will be low if the reaction is stopped before 1 hour.

ix. After 1 hour, the heating mantle is separated, and the reaction liquor is disseminated into centrifuge tubes (each containing 5–10 mL of reaction liquid). Hexane is added to the centrifuge tube until each tube has a total of 20–25 mL solution. This step is critical because the precipitation of the quantum dots directly from the reaction liquid will result in oil in place of solid precipitates after centrifugation process. Ethanol is then incorporated into the mixture until the color of transparent yellow solution changes to opaque (25 mL). The centrifugation of mixture is carried out at 7500 rpm for 300 seconds. The centrifugation speed and time is not critical, as far as the resultant product is a solid precipitate (mostly colorless). If the final product is not in the form of solid precipitates, the use of higher hexane/liquor ratio (or lower solvent/antisolvent ratio) is recommended (Thon et al., 2013; Zhitomirsky et al., 2014).

x. The CdS quantum dots are dispersed in 9 mL volume of toluene. It is essential to wash the products only a single time to avoid the gelation of QDs in toluene which results in difficulties in the injection of the mixture. This reaction consistently generates QDs having the first exciton peak at 366 nm (Kim et al., 2015; Xu et al., 2017).

3.4.3. Synthesis of CdSe QDs

i. The precursor solution is developed by mixing 6.28 g of OA, 25 g of the ODE and 0.512 g of CdO followed by heating in the similar pattern as in case of CdS QDs synthesis steps 1 and 2.

ii. During the heating of reagents, 0.8 mmol (0.063 g) of Se powder is incorporated to 5 mL of ODE followed by sonication for 10 minutes. Additionally, 3 mL of 1 M solution of TOPSe is blended with 7 mL of ODE succeeded by loading into a syringe pump.
   a. The sonication time is not critical provided that the mixture is vigorously sonicated immediately before loading of the injection syringe. The Se particles begin to settle out of the ODE solution if left for a prolonged duration. The recommended sonication time is at least 10 minutes to disperse the Selenium powder well; however, the sonication time can increase further as far as the Se particles do not settle before injection.
   b. The TOPSe solution is synthesized by stirring 0.1 mol of Selenium powder in 100 mL solvent on TOP in a glovebox for approximately 3 hours. All of the Se powder must be dispersed completely to form a transparent (or yellow-tinted) solution before use in the synthesis process. If TOPSe is stored in the glovebox, it can be utilized for an extended period. A clear yellow color or precipitation at the bottle-bottom is an indication of the TOPSe oxidation which results in discarded TOPSe.

iii. The injection of ODE/Se mixture takes place at 260°C, and the temperature control is set to the temperature of 240°C.

iv. After 1 minutes from the injection, the ODE/TOPSe solution is incorporated dropwise at 1 drop/second. This pattern is set by naked-eye using the syringe pump. This step can also be performed by hands, but a better consistency is achieved with a syringe pump. The injection of the ODE/TOPSe solution by hand requires the process to be as steady as possible to enhance the size distribution of the resultant CdSe.

v. During the addition of ODE/TOPSe, the UV−vis absorbance reactions are monitored until the desired size distribution is achieved. This step is most conveniently performed by drawing small aliquots (around 0.1 mL) which are quenched in a cuvette containing hexane (around 2 mL). Aliquots should be observed every 5−10 minutes to track the QDs growth.

vi. The addition of ODE/TOPSe mixture is ceased when the first exciton peak of the CdSe QD reaches 600 nm following by turning-off the heating mantle 2 minutes after the stoppage of ODE/TOPSe mixture injection, allowing the quantum dots to cool to ambient temperature. On average, around 3 mL of the ODE/TOPSe mixture is left and the total time of reaction around 30 minutes. This time varies depending on the speed of dripping. The slower additions of ODE/TOPSe result in higher monodispersity in the final product.

vii. The reaction liquor gets split into 50 mL centrifuge tubes in a way that there is approximately 5 mL liquor in each tube. Around 20 mL of hexane is supplemented to each tube followed by capping and shaking of the tube. 25 mL of ethanol also added to each tube succeeded by the centrifugation of tubes at 7500 rpm for 5 minutes. Typically, this first precipitation yields slightly colored supernatant which is discarded while preserving the precipitated QDs (Sayevich et al., 2015).

viii. The washing of QD product is carried out twice by dispersing each precipitate of QD in 10 mL liquid of hexane followed by the precipitation in an equal ethanol volume. The quantum dot product is disseminated in 9 mL of ODE after the final wash. Dispersion of the QDs requires the assistance of a vortex mixer. The storage of the CdSe QDs mixture for a long time necessitates their transportation is a glovebox before their final dispersion in an ODE.

ix. A product of CdSe quantum dots with narrow size distribution exhibits a sharp peak of first exciton and a clear feature of the second exciton as depicted in the Figure 3.6.

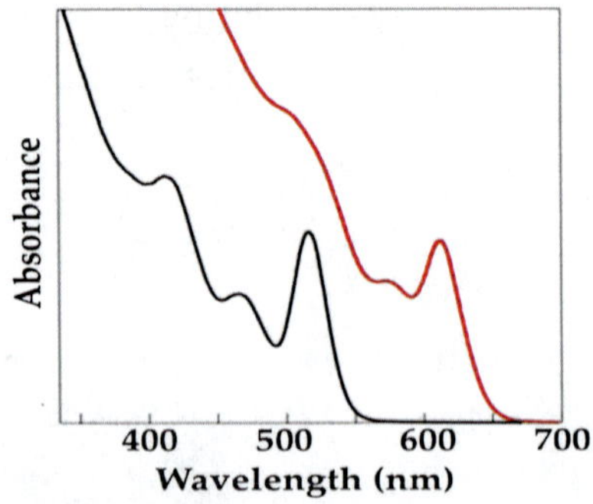

Figure 3.6 Absorbance spectra of CdSe quantum dots with the first exciton peak at 610 nm (red) and 515 nm (black). [*Source*: https://pubs.acs.org/doi/abs/10.1021/acs.chemmater.6b02939].

3.4.4. Cation Exchange Reactions from CdSe/CdS to PbSe/PbS

PbS/PbS QDs are fabricated by a cation exchange reaction using CdS/CdSe quantum dots.

- The precursor solution is developed by mixing 3 mmol (0.834 g) of $PbCl_2$ and 3 mmol (10 mL) of OLA in a round-bottomed flask of 100 mL capacity.
- The mixture is exposed to a thermal treatment at 100°C with continuous stirring and degassing using other environments, i.e., vacuum and N_2.

Figure 3.7. (a) Heating of the OLA/PbC12 mixture takes place when the mixture shows a clear appearance. (b) Formation of milky white solution results after heating the OLA/PbC12 mixture to about 140°C. [*Source:* https://pubs.acs.org/doi/abs/10.1021/acs.chemmater.6b02939].

i. At a temperature of 100°C, a needle is pierced into 1 septum, as displayed in the above figure, to instigate the flow of N2 over the reagents for the duration of approximately 5 minutes. This practice is used to remove water molecules further previously not removed by degassing.

ii. The mixture is exposed to a temperature of 140°C followed by the removal of the vent needle. Heating of the suspension results in the formation of a complex from OLA and PbCl2 followed by the clear appearance of the solution (Figure 3.7a). At temperatures above 140°C, the solution turns turbid and milky again (Figure 3.7b). The temperature of the solution is retained at 140°C for 10 minutes. The hold of 10 minutes is not critical as far as the milky white solution is attained before the injection of CdS (CdSe) quantum dots (McGuire et al., 2008; Nozik, 2008). Formation of the OLA/$PbCl_2$ complex displays some color differences at this instance of the reaction. The solution normally remains clear until agitated to higher temperatures (above 145°C) or the solution exhibits a pink appearance after the formation of the complex. Neither of the cases negatively impacts the synthesis process (Moreels et al., 2008; Fritzinger et al., 2010).

iii. Heating or cooling of the reaction bath up-to-the injection temperature (195°C to fabricate 0.7 eV PbSe quantum dots and 90°C to produce 1.3 eV PbS quantum dots) is carried out to initiate the quantum dot synthesis. Injection of 2–3 mL of CdS or CdSe quantum dots (capped with cadmium–oleate, 50–100 mg/mL in an ODE) is carried out using a 16-gauge needle and 20 mL syringe. Roughly the injection of 200 mg total of CdS or CdSe QDs is required. Injection process renders the solution color to dark brown which is a consequence of the cation exchange reaction. The large diameter needle and syringe are essential to ensure a rapid injection of the CdS or CdSe quantum dots. The resultant Pb QDs usually have a wide size distribution if the injection of CdS (CdSe) QDs is not carried out swiftly enough (Bae et al., 2012; Zhang et al., 2013).

iv. The reaction is permitted to take place for 30–60 seconds (contingent to the desired size) before removal of the mantle followed by immediate quenching by immersing the flask in an aqueous bath. Lower reaction temperature coupled with shorter reaction time may result in the synthesis of QD blend containing a mixture of PbS, PbSe, CdS, and CdSe QDs (22–23).

v. Injection of 10 mL of hexane is carried out at 70°C to facilitate cooling. Around 8 mL of OA is injected at 30°C to substitute the weakly bound ligands of OLA on the surfaces of the PbS or PbSe QDs. The injection of OA results in the rise of solution temperature to 45°C which follows the cooling of the mixture below 30°C before exposure of the solution to the atmosphere (Selinsky et al., 2012; 2013).

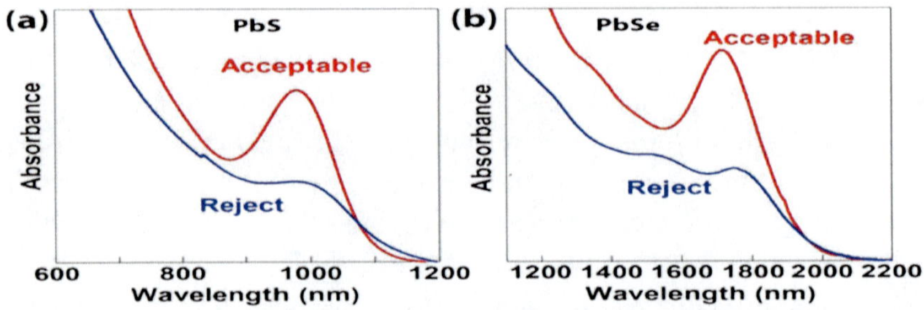

Figure 3.8. Examples of (blue) unacceptable (red) acceptable absorbance spectra (absorbance peaks) of (a) PbS QDs and (b) PbSe QDs [*Source*: https://pubs.acs.org/doi/abs/10.1021/acs.chemmater.6b02939].

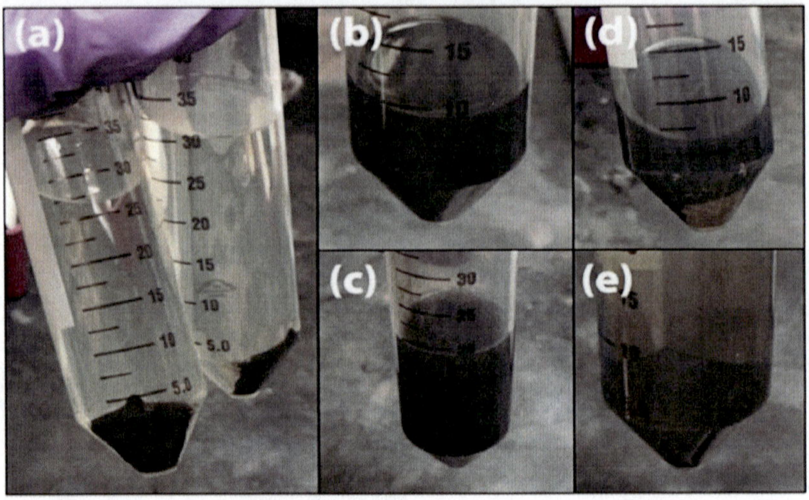

Figure 3.9. Washing process for QDs with cation exchange. (a) Formation of the precipitates at the centrifuge tube bottom after precipitation and centrifugation process; (b) fully dispersed QDs in hexane; (c) cloudy, gray solution which consequences from the formation of precipitates due to ethanol; (d) white precipitation products of surplus chlorides after centrifugation process carried out in neat hexane solvent; (e) Finishing precipitation and centrifugation. It should be noted that some QDs are stuck to tube walls present above the main solid

precipitates, which indicates the little washing of QDs. [*Source*: https://pubs.acs.org/doi/abs/10.1021/acs.chemmater.6b02939].

vi. The cooled reaction product is inserted into 2 centrifuge tubes of 50 mL which are evenly split to accommodate 15 mL of reaction products. Additional hexane (around 5 mL) is employed to rinse the flask (round-bottom), and this is also incorporated into the centrifuge tubes. Addition of 20 mL ethanol is carried out to initiate the precipitation of the QDs (Schaller & Klimov, 2004; Hanna et al., 2012; Schaller et al., 2015).

vii. Centrifugation of the mixture is carried out at 7500 rpm for 5 minutes.

viii. The colorless resultant supernatant is typically discarded (Figure 3.8a). If the supernatant exhibits a darker color, additional ethanol (2–4 mL) is added followed by centrifugation.

ix. The solid quantum dot precipitates in the tube are typically dissolved in 10 mL hexane solvent. The solutions are united into 1 centrifugal tube followed by the centrifugation of the tube without incorporation of an anti-solvent. This particular step instigates the precipitation (white precipitates) of the excess chlorides in the solution.

x. A syringe is used to extract the QD solution followed by filtering of the solution using a 0.2 μm filter. The latter step facilitates the removal of excess chloride salts .(Figure 3.9)

xi. A final wash is performed by carrying out precipitation using 20 mL of ethanol followed by centrifugation.

xii. For the dip-coating process, dispersion of quantum dots in hexane takes place at 10–20 mg/mL. For spin-coating technique, dispersion of quantum dots in octane takes place at 30 mg/mL.

3.5. QD CHARACTERIZATION

After every fabrication process, the quantum dots are characterized using UV–vis-NIR spectroscopy. The absorbance and transmittance spectra are utilized to determine the size, concentration, and polydispersity of quantum dots. In general, the impact of polydispersity is to restrict the V_{oc} because of the enhanced energy disorder. Polydispersity is measured via the breadth of the initial exciton peak. Figure 3.10 illustrates an example of polydispersity levels (acceptable or unacceptable level) for PbS and PbSe QDs. The polydisperse quantum dot batch must not be employed, or the precipitation with controlled size-selection should be achieved to isolate monodisperse QDs (Sitt et al., 2013; Wu et al., 2015).

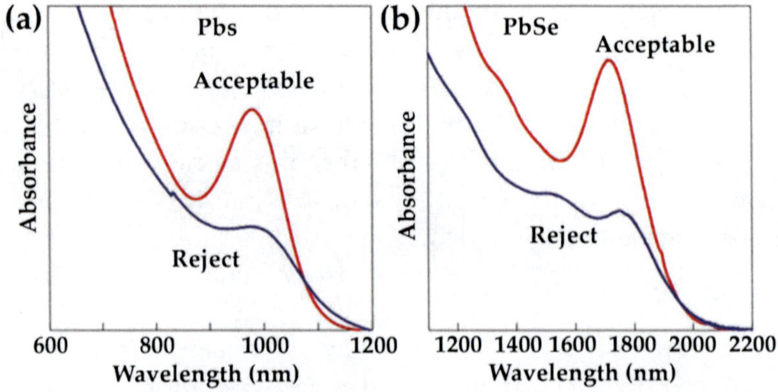

Figure 3.10. Examples of (blue) unacceptable and (red) acceptable absorbance spectra of (a) PbSe and (b) PbS QDs [*Source*: https://pubs.acs.org/doi/abs/10.1021/acs.chemmater.6b02939].

CHAPTER 4

NITROGEN-DOPED CARBON QDS FOR GREEN PHOTOVOLTAIC CELLS

CONTENTS

4.1. BACKGROUND

Taking into consideration the significant factor of the environment protection, now, "green" materials are explored for photovoltaics. In this chapter, we explained the development of a type of quantum dots solar cell by making use of nitrogen-doped carbon dots. These nitrogen-doped carbon dots have been prepared through direct pyrolysis of ammonia and citric acid. (Ip et al., 2012; Sargent, 2012) The excitonic absorption of these nitrogen-doped carbon dots' is dependent on the content of N-doping in the carbon dots. This N-doping can be altered readily by the mass ratio of reactants. The resultantly created "green" nitrogen-doped carbon dots solar cell accomplishes the best power conversion efficiency of around 0.79% under AM 1.5 G one full illumination of sun, which so far, is the highest efficiency for any carbon dot-based solar cells (Sambur et al., 2010; Semonin et al., 2011; Kim & Ma, 2014).

Carbon quantum dots (known as CDs) are an interesting sub-class of carbon nanoparticles. Carbon nanoparticles are under investigation for numerous applications owing to their favorable low toxicity, optical stability, ease of functionalization, and biocompatibility (Kramer & Sargent, 2013; Chuang et al., 2014; Du et al., 2016). Several researchers have investigated the synthesis methods as well as photoluminescence properties of CDs. Till now, multiple approaches of synthesis such as electrochemical exfoliation, laser ablation, incomplete combustion oxidation, pyrolysis, microwave synthesis, acidic oxidation, and hydrothermal treatments have been developed for preparing CDs (Pan et al., 2013; Santra & Kamat, 2013; Zhu et al., 2013). Many raw materials such as citric acid, graphite oxide, glycerol, soy milk, eggs, coffee grounds, and grass have been used for the synthesis of CDs (Wang et al., 2013; Ji et al., 2015; Jiao et al., 2015). Nevertheless, it is still desired to be able to rapidly synthesize high-quality CDs through a feasible and environmentally friendly method consuming readily available and low-cost raw materials (Zhang et al., 2013; 2014; Speirs et al., 2015). An example of such suitable raw material is lignin. Lignin is an abundantly occurring natural organic polymer and is a rich source of carbon (Neo et al., 2014; Pan et al., 2014; Lim et al., 2015). An increasing interest has been developing for the use of lignin to prepare novel carbon-based materials (Markovic et al., 2012; Ge et al., 2014). Still, it is not easy to degrade lignin and its derivatives owing to their molecular structure which contains active carbon to carbon linkages. Among several techniques, the process known as hydrothermal carbonization is so far a promising method for the synthesis of innovative carbon-based materials, particularly CDs (Hu et al., 2009; Baker & Baker, 2010).

In the class of third-generation photovoltaics, a promising emerging candidate is quantum dot solar cells (QDSCs) which have been increasingly used owing to their versatile and unique characteristics of quantum dots (QDs) including high absorption coefficient and tunable band gap (Zhou et al., 2007; Zhao et al., 2008; Liu et al., 2009). Normally, low-bandgap metal chalcogenide (such as $CuInS_2$, CdTe, CdS, PbS, CdSe, PbSe, etc.) QDs have been extensively employed as sensitizers in QDSCs (Ding et al., 2013; Ye et al., 2013). Although QDSCs have been well-developed, still they comprise of highly toxic metals (such as In, Cd, and Pb). Henceforth, there is extreme need of environment-friendly alternatives ("green" materials), and they are welcomed for the fabrication of solar cells (Wang et al., 2014; 2015).

Due to the unique electronic and optical properties, excellent biocompatibility, water solubility, robust chemical inertness, and low toxicity (Zheng et al., 2014; Zhu et al., 2014; Cuesta et al., 2015), carbon dots (CDs) have been extensively researched as well as applied in various fields as a promising carbon-based nano-material (Li et al., 2010; Tavallali et al., 2011). One of the applications of carbon dots is its use as sensitizers for QDSCs (Sun et al., 2006; Guo et al., 2012). For example, Mirtchev et al. (2012) have developed water-soluble CDs to be used as sensitizers for QDSCs accomplishing power conversion efficiency (PCE) of 0.13%. On the other hand, Zhang et al. (2013) have fabricated 0.13% PCE of QDSCs that are based on nitrogen-doped CDs (Nano Carbon QDs). Till now, the main factor which limits the efficiency of cells might be the 'low short-circuit current-density (J_{sc})' in contrast with the fill factor (FF) and open-circuit voltage (V_{oc}) of metal chalcogenide-based QDSCs. Diverse trap states and narrow absorption of light lead to extremely low J_{sc} of CD-based QDSCs. In the meantime, CDs' light-absorption is very weak at visible region, and it is centered at the ultra-violet region. Since the subsistence of trap state defects with different levels of energy has been established through their excitation-dependent fluorescence, therefore, moderately photo-induced carriers will be recombined by CDs instead of being injected into the transporting media. Consequently, the PCE of QDSCs is extremely low. There needs considerable effort for improving the possible sensitizers for QDSCs (Mao et al., 2014).

In this chapter, the development of Nano Carbon QDs has been explained for building up the "green" QDSCs with improved efficiency (Yu et al., 2012; Tian et al., 2015). These Nano Carbon QDs were developed by direct pyrolysis of ammonia and citric acid. Scientists have conducted different experiments, the results of which exhibits that the Nano Carbon QDs' excitonic absorption is mainly dependant on the N-doping content altered by the mass ratio of reactants. The Nano Carbon QDs are having the best excitonic absorption have been used as sensitizers in the QDSCs (Yan et al., 2010; Briscoe et al., 2015).

4.2. SYNTHESIS OF CARBON QDS

Direct pyrolysis method has been extensively used to prepare the Nano Carbon QDs. The preparation of Nano Carbon QDs is usually carried out by direct pyrolysis of ammonia (Sinopharm Chemical Reagent Co) and citric acid (CA, Alfa Aesar) mixture, just like the previous methods suggested by different researchers (Dong et al., 2012; Peng et al., 2012). Normally, CA (4 g) and a specific quantity of ammonia are dissolved in distilled water (around ten mL) and kept stirring for almost 30 min. After that, we transfer the prepared solution into a porcelain boat followed by heating it in the air at 200°C for almost 3 hours with a heating rate of 10°C/min. The acquired carbogenic product is then mixed with acetone (40 mL) via ultrasonic. After 15 minutes of centrifugation at 10,000 rpm, the supernatants that contain Nano Carbon QDs are collected. Afterwards, rotary evaporation is employed to remove the acetone, and solid-state Nano Carbon QDs are gathered for further characterization as well as usage (Sun et al., 2014; Li et al., 2014; 2015).

4.2.1. Fast Synthesis of Carbon QDs Through Hydrothermal Lignin Treatment

In this section, we have explained a rapid method for the synthesis of highly luminescent CDs through the hydrothermal treatment of lignin with the support of H_2O_2. It is a common fact that H_2O_2 can be easily dissociated into hydroxyl radicals in the presence of the photo-assisted catalysis Fe^{3+}/Fe^{2+} in water. The resultant OH radical is an enormously powerful oxidizing species (Kwon et al., 201; Wu et al., 2013). This approach for the synthesis of Carbon Quantum Dot is simple as well as environment-friendly. It has been demonstrated that the prepared CDs exhibit good water solubility, good luminescence property, and low cytotoxicity, and narrow particle size distribution. The CDs also exhibited excellent bio-imaging capacities in Hela cells. This work has provided a novel approach for the synthesis of CDs from natural materials. Moreover, it demonstrates the capabilities of CDs in bio-imaging applications (Armel et al., 2011; Li et al., 2012).

4.2.2. Hydrothermal Treatment of Lignin

Figure 4.1 displays the pyrolysis products of lignin at different times (starting from 10 min till 60 min having gap size of 10 min in between). It can be seen that as the time increases, the color of the initial product solution turns pale (Li et al., 2011; Sahu et al., 2012; Tetsuka et al., 2012).

The residual lignin is kept on the filter paper (see the Figure 4.1b). The synthesis outcome of the CDs has been observed as 12.06%, 10.2%, 6.67%, 2.3%, 1.05%, and 0.8% in proportion to 10, 20, 30, 40, 50 and 60 min, respectively. The obtained optical images of the CDs solution display blue luminescence under UV light excitation.

Maximum brightness is usually observed in the case when the pyrolysis time is 40 min. The CDs (carbon QDs) solution continues to remain transparent for almost half a year without precipitation.

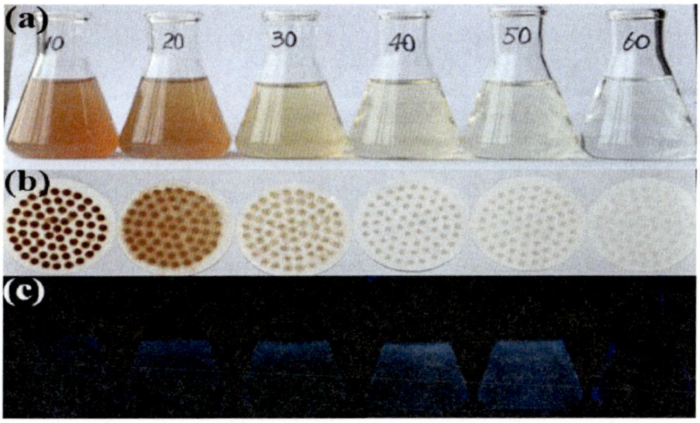

Figure 4.1. Results of Pyrolysis for (a) the initial product solution; and (b) the residual lignin after 10, 20, 30, 40, 50 and 60 min of hydrothermal reaction; and (c) the resulting filtrate image under 365 nm UV lamp irradiation [*Source*: https://www.ncbi.nlm.nih.gov/pmc/articles/PMC5456677/].

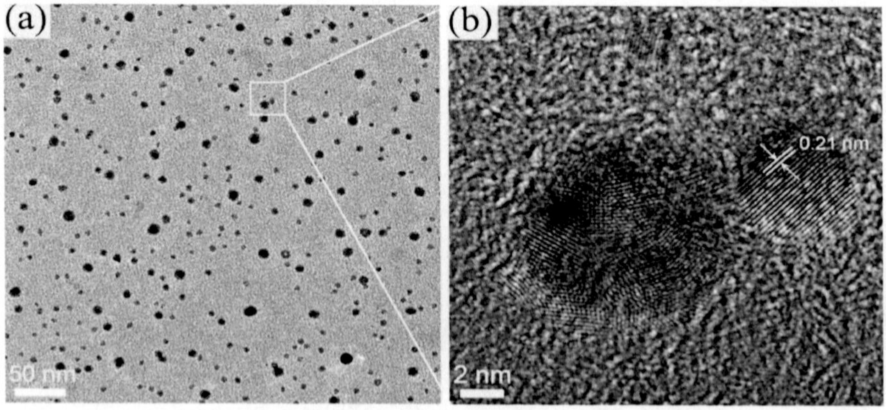

Figure 4.2. (a) Transmission electron microscopy (TEM); (b) high-resolution TEM (HRTEM) photographs of the carbon dots (CDs) synthesized by hydrothermal treatment of lignin [Source: https://www.ncbi.nlm.nih.gov/pmc/articles/PMC5456677/].

High-resolution TEM (HRTEM) and Transmission electron microscopy (TEM) is performed on the CDs, and the results have been displayed in Figure 4.2. The size of the CDs has ranged from 2 to 10 nm. The HRTEM images show that the carbon dots have crystalline structure; moreover, the distance of lattice spacing is around 0.21 nm, near to that of the graphite (100) plane (Cao et al., 2007; Yang et al., 2008).

Figure 4.3 shows the wide-scan X-ray photoelectron spectroscopy (XPS) spectrum. In this image, 2 strong peaks were attributed to carbon and oxygen at 532.0 and 285.5 eV, respectively. The elemental constituents of the as-prepared CDs were O (17.42%) and C (82.58%). The de-convoluted C_{1s} spectrum displayed three constituents, which could be assigned as, sp^3 carbon at ~286.1 eV, graphite (sp^2) carbon at ~283.2 eV and carboxyl carbon at 288.6 eV. The O_{1s} spectrum (Figure 4.2c) displays three peaks at 530.6, 532.1 and 533.2 eV, which are ascribed to the C=O, C-OH and C-O-C groups, in that order (Sun et al., 2006; 2008; Anilkumar et al., 2011).

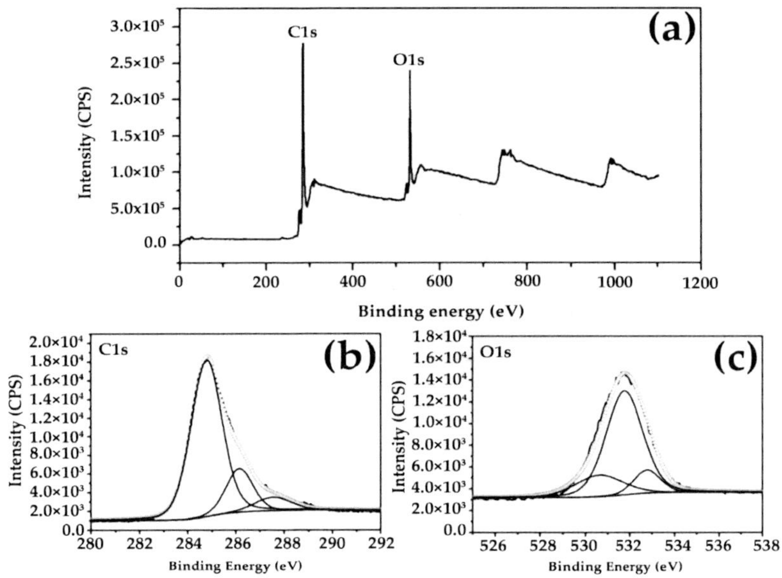

Figure 4.3. (a) X-ray photoelectron spectroscopy (XPS); (b) C1s; (c) O1s spectrum of the as-prepared CDs [*Source*: https://www.ncbi.nlm.nih.gov/pmc/articles/PMC5456677/].

The CDs obtained after 40 min of pyrolysis are characterized using Raman spectroscopy, as can be seen in (Figure 4.4). The resultant Raman spectrum displayed a strong D band at 1382 cm^{-1}, which corresponds to the sp^3 faults in CDs. Besides, at 1578 cm^{-1}, a G band is observed, which corresponds fine with the disordered carbon as well as the sp^2 clusters, signifying that there are aromatic groups within CDs. It has experimented that the CDs have an I_D/I_G ration of around 0.91, which may be because of oxygen-rich edges of the CDs. The UV-Vis spectrum of an aqueous solution of CDs exhibits 2 peaks at 282, and 348 nm, representing that various surface states are present in the CDs solution. The measurement of fluorescence spectra of the CDs is usually carried out with an F-4500 fluorescence spectrometer (HITACHI, Tokyo, Japan), having a slit whose width was 10 nm for both excitations as well as emission beams (Xu et al., 2004; Cai et al., 2011; 2012). The excitation wavelength shows continuous alteration from 280 to 500 nm, in regular increments of 20 nm. Figure 4.4b displays the corresponding spectra. Colorful and Bright photoluminescence (PL) emissions had been observed from the CDs. With the increase in the excitation wavelength, the emission maxima shifted and displayed the highest PL intensity at an emission wavelength of 430 nm and an excitation wavelength of 320 nm (Jia et al., 2012; Liu et al., 2012).

The CDs exhibit excellent photo-stability since the fluorescence intensity does not vary, even after constant excitation under a 150 W Xenon lamp. The fluorescence of the fluorescein isothiocyanate (FITC) shows rapidly reduction within few minutes of excitation. Further, the CdTe quantum dots (QDs) conserve 25% of the original PL intensity after excitation of around 20 min. This has been displayed in Figure 4.4. These results have indicated that the PL of the CDs is quite more stable as compared to the fluorescent FITC as well as the CdTe QDs. It is assumed that the construction of CDs, as well as their surface functionalization, takes place concurrently during the process of hydrothermal carbonization. The existence of a huge number of carboxylic acids introduces many different surface defects. Such defects behave like excitation energy traps and are liable for the different photoluminescence behaviors. Quite a few mechanisms have been planned to explain these distinctive PL properties, for instance, distribution of different emissive trap sites, the size distribution of the CDs, and the development of numerous different polyaromatic fluorophores within the carbogenic core. Nevertheless, the precise mechanism of the behavior of CDs' PL is still somewhat unclear, and further studies are essential to understanding this property in detail (Rahim et al., 2007) (Figure 4.5).

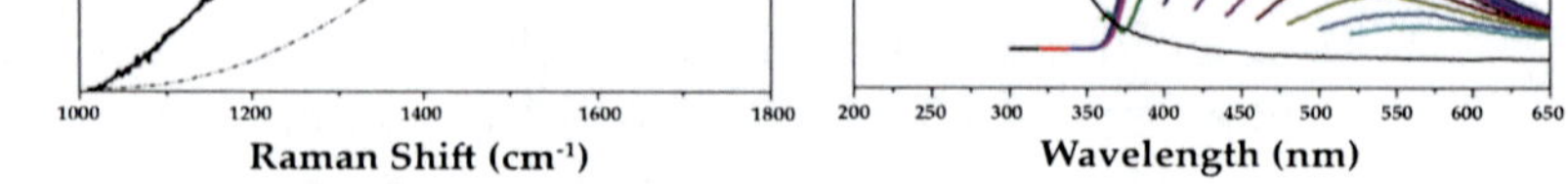

Figure 4.4. (a) Raman absorbance; the background is denoted by the dashed line; (b) photoluminescence (PL) spectra of the CDs at varying excitation wavelengths [*Source*: https://www.ncbi.nlm.nih.gov/pmc/articles/PMC5456677/].

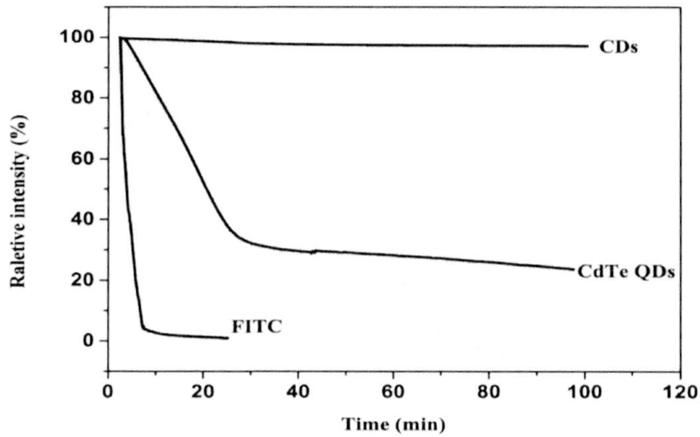

Figure 4.5. Comparison of photo-stability of the CdTe QDs, fluorescent CDs, and fluorescein isothiocyanate (FITC) in a fluorescence spectrophotometer using a 150 W Xe lamp under 360 nm excitation. [*Source*: https://www.ncbi.nlm.nih.gov/pmc/articles/PMC5456677/].

For investigating the feasibility of employing CDs for use in bio-imaging, A549 human lung adenocarcinoma cells are usually employed for evaluation of the cytocompatibility of the CDs. Using a methylthiazoleterazolium (MTT) assay, the cell feasibility of the CDs can be determined. As can be seen in the Figure 4.6, the MTT assays of cell feasibility reports depict very low cytotoxicity of the CDs. This result endorses that CDs can be employed for imaging or various other biomedical applications (Dutta, 2000; Wang et al., 2010).

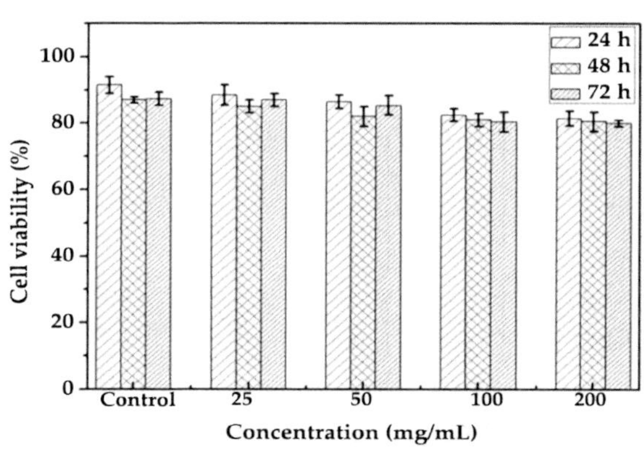

Figure 4.6. Cytotoxicity evaluations test of A549 cells having different amount of CDs40min after 24, 48 and 72 hours of incubation [*Source*: https://www.ncbi.nlm.nih.gov/pmc/articles/PMC5456677/].

After attaining CDs from 40 min hydrothermal carbonization, they are introduced into the Hela cells, after which their bio-imaging abilities are assessed using *in vitro* confocal microscopy test. The previous results demonstrated that the photo-luminescent spots were detected merely in the cell membrane and cytoplasmic zone of the cell, which indicates that the CDs were capable of penetrating into the cell effortlessly. This observation made is following earlier studies on the interaction between living cells and nanomaterials. These results prove that CDs can be employed as a fluorescence probe for bio-imaging applications (Bourlinos et al., 2008; Yang et al., 2009). (Figure 4.7).

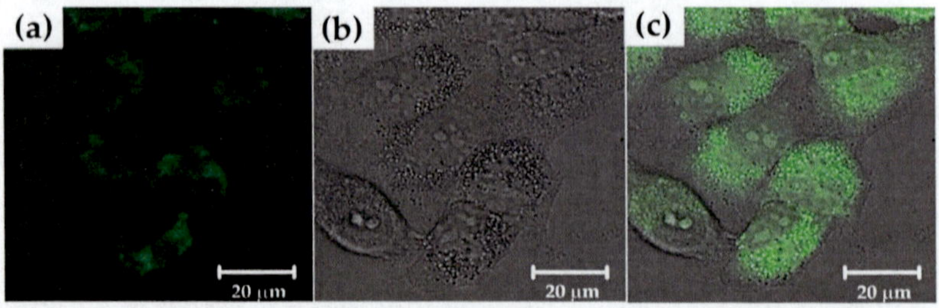

Figure 4.7. (a) A confocal fluorescence micro image of Hela cells written off with the CDs (λex: 405 nm); (b) A bright field micro image of the cells; (c) An overlap image of (a) and (b) [*Source*: https://www.ncbi.nlm.nih.gov/pmc/articles/PMC5456677/]

4.3. SYNTHESIS AND CHARACTERIZATION OF CARBON QDS

4.3.1. Preparing the Fluorescent CDs

In a usual procedure, synthesis of fluorescent CDs is carried out in the following manner:

100 mg of lignin is dispersed in sanitized water (around 30 mL) and after that ultra-sonication takes place for 10 minutes followed by the addition of 2 mL H_2O_2 in the solution, and finally, the mixture is made airtight in a 50 mL Teflon crinkled stainless steel autoclave. This autoclave is afterward placed in a muffle furnace, subsequently applying the hydrothermal treatment on it at 180 degrees Celsius for 10, 20, 30, 40, 50 and 60 minutes. Next, to the reaction, natural cooling of the autoclave takes place, and the yellow solution obtained as a result is filtered with a 0.22 μm membrane filter for

removing the unreacted lignin. This filtrate is then exposed to dialysis for 2 days by employing a 3500 Da dialysis membrane for removal of the excess H_2O_2. To obtain the final CDs, the resultant yellow solution is freeze-dried (Ray et al., 2009; Xie et al., 2012).

4.3.2. Characterization Methods

Morphological characteristics of the CDs are investigated using field emission electron microscope (JEOL JEM-2100F, JEOL, Tokyo, Japan) and transmission electron microscopy (TEM). X-ray photoelectron spectroscopy techniques (XPS, AXIS ULTRA DLD) are employed for investigation of the functional groups existing on the surface of the CDs. The measurement of fluorescence spectra of the CDs is carried out by using a fluorescence spectrometer (F-4500, HITACHI, Tokyo, Japan), having a slit whose width was 10 nm for both emissions as well as excitation. The wavelength of excitation showed an increment by a 20 nm initiating from 280 to 500 nm (Zhang et al., 2012).

4.3.3. Fluorescence Imaging Experiments

Hela cells are planted in each well of a confocal dish and cultured at 37 degrees C for 24 hours. An aqueous mixture of the CDs (0.1 mg/mL) is passed over a 0.2 μm disinfected membrane filter. The resultant filtered fluorescent suspension (40, 50, and 60 μL) is then mixed with the culture medium (200 μL) and afterward incorporated with 3 wells of the confocal dish (the fourth one was used as a control) where the Hela cells are grown. The medium is removed after an incubation period of 2 hours, and the cells are thoroughly washed 3 times by phosphate buffered saline (PBS) and kept in it for optical imaging. CD's cellular uptake by Hela cells is normally observed through confocal microscopy using the excitation wavelength of 405 nm.

4.4. FABRICATION OF SOLAR CELL

The fabrication of Nano Carbon QD-sensitized TiO_2 photo-anodes is carried out according to the method suggested by Li et al. (2012). The washed fluorine-doped tin oxide (FTO) glasses are submerged in 40 mM $TiCl_4$ aqueous solution for 30 min at 70°C and then washed with ethanol and water. A TiO_2 paste having a size of 20 nm is deposited on the FTO glass plate through doctor blade printing method and then dried for 5 min at 125°C. The smattering layer of a TiO_2 paste having 200 nm size is coated onto the top of the first layer of TiO_2 after which it is sintered in air for 30 min at 500°C. Next, to cool down to 95°C, the TiO_2 film is submerged in the NCQD acetone solution for 2 hours at room temperature. The Nano Carbon QD sensitized TiO_2 is cleaned using anhydrous ethanol and then dried out with a nitrogen stream. The fabrication of solar cells takes place by sandwiching gel electrolytes in between a Pt counter electrode and a NANO CARBON QD-sensitized TiO_2 electrode. These electrodes are separated by a hot-melt ring having a thickness of 25 mm and sealed through heating. Finally, the electrolyte injection hole existing on the thermally Platonized FTO counter electrode is sealed using a thin glass and Surlyn sheet through heating (Qu et al., 2013; Zhao et al., 2014).

4.5. CHARACTERIZATION AND MEASUREMENT

The investigation of the morphology of the materials is done through atomic force microscopy and transmission electron microscopy. Scientists have obtained Fourier transform infrared (FT-IR) spectra on a Bruker VERTEX70 FT-IR spectrometer whose range is from 4000 to 400 cm^{-1} whereas, X-ray photoelectron spectra (XPS) has been attained using a Japanese Kratos Axis Ultra HAS spectrometer employing a unicolor Al Kα source. The characterization of crystallinity of Nano Carbon QDs can be carried out via Raman spectroscopy (HORIBA Jobin Yvon HR800) by employing 514-nm laser as the source of excitation. The recording of UV–visible absorption spectrum, as well as diffuse-reflectance spectra, can be done on a UV2501PC (Shimadzu); whereas, the photoluminescence (PL) spectra can be recorded on a Fluoromax-4 spectro-fluoro-meter (HORIBA JobinYvon Inc) which contains a 150 Watt of a xenon lamp as the source of excitation. Scientists have used a solar simulator for the measurement of photocurrent-voltage with an AM 1.5 G filter using an irradiation intensity of about 100 $mWcm^{-2}$. The calibration of light intensity can be done using a conventional silicon photovoltaic solar cell. The active area of the cell is 0.25 cm^2. All of the above mentioned measurements are executed at room temperature (De & Karak, 2013).

4.5.1. UV-VISIBLE SPECTROSCOPY

The UV–visible spectra of Nano Carbon QDs from various reactant mass ratios exhibits that Nano Carbon QDs' excitonic absorption is mainly dependent on their N-doping content (Figure 4.8). The results acquired establishes that the N-doping content can be effortlessly altered via the mass ratio of reactants.

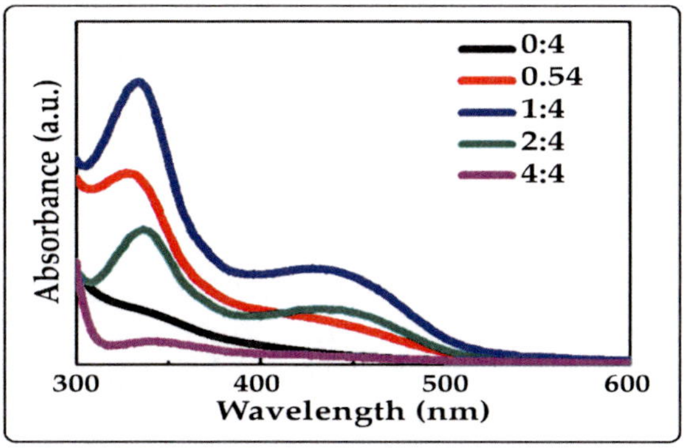

Figure 4.8. UV–vis spectra of Nanocarbon dots from 4 g CA and several quantities (0, 0.5, 1, 2 g) of ammonia [*Source*: https://link.springer.com/article/10.1186/s11671-016-1231-1].

When the mass ratio of the reactants CA and ammonia and CA is around 4:1, the Nano Carbon QDs have the best visible absorption. The surplus ammonia may excite a strong chemical reaction with inflamed Nano Carbon QDs that can result in the extreme amount of water loss as well as damage to the surface of Nano Carbon QDs (Prasannan & Imae, 2013).

4.5.2. X-Ray Photo-Spectroscopy (XPS)

To investigate the impact of the mass ratio on the levels of N-doping in the Nano Carbon QDs, the elementary constituents of the Nano Carbon QDs have been evaluated in the Table 4.1. By increasing the mass ratio of CA and ammonia, the N-doping content increases. However, it changes slightly when there is further increase in mass ratio further over 4:1 (Table 4.1).

Table 4.1. Comparative Elemental Analysis of The Carbon Quantum Dots with Different N-Doping Levels Through XPS Analysis

Elemental/atomic (%)	**N**	**O**	**C**
N-free	0	21.6	78.4
0:4	2.6	22.6	74.8
1:4	10.0	22.5	67.5
2:4	9.7	21.1	69.7
4:4	9.3	19.2	71.5

Again, the optical features of optimal Nano Carbon QDs are studied in detail. As compared to N-free CDs, the finest Nano Carbon QDs display very higher absorption peak around 335 nm; moreover, they also have more wide absorption band which extends to 550 nm (Figure 4.9). The absorption peak near 335 nm correlates with n–π* transition of C=O bonds whereas; the visible absorption band is accredited to the amino groups on the surface of Nano Carbon QDs (Zhou et al., 2012; Yan et al., 2014). The increasing absorption up to 550 nm explains nitrogen-doping CDs in some way. Additionally, Nano Carbon QDs also demonstrate excellent solubility with dazzling blue emission under UV lamp (wavelength of 365 nm) irradiation. With the wavelength of excitation rising from 300 to 500 nm as shown in Figure 4.9, the intensity of emission decreases gradually and hence the emission peak shifts from 425 nm to 510 nm (Schmidbaur & Schier, 2012; Fu et al., 2013). The surface groups of Nano Carbon QDs bring in trap states having different levels of energy that result in the excitation-dependent emission. The Nano Carbon QDs' fluorescence quantum yield (FLQY) when calculated is ~36% with reference being the quinine sulfate (FLQY 54%).

4.5.3. Interpretation of Characterization Data

The characterization of Nano Carbon QDs is also done by FT-IR spectroscopy. C–N stretching vibrations absorption bands at 1182 cm^{-1} verify the doping of N in the Nano Carbon QDs. Moreover, O–H stretching vibrations have also been observed at 3000–3400 cm^{-1}, C=O stretching vibrations at 1710 cm^{-1}, C–H stretching vibrations at 2940 cm^{-1}, C–O–C stretching vibrations at 1350 cm^{-1}, and C–O stretching vibrations at 1286 and 1408 cm^{-1}. XPS measurements are carried out for analysis of the surface element of Nano Carbon QDs as can be seen in the Figure 4.9.

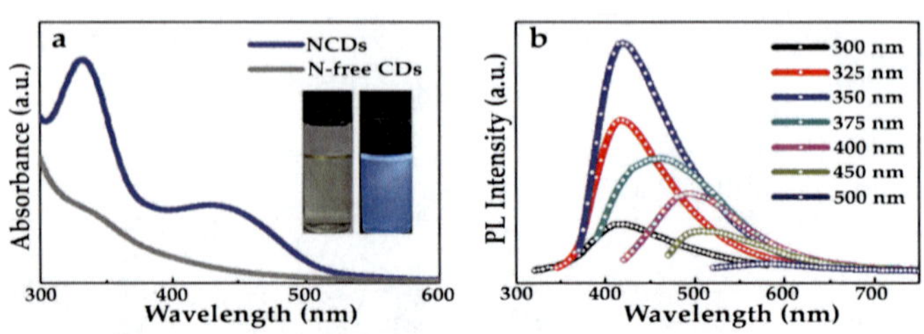

Figure 4.9. (a) UV–visible spectra of nanocarbon dots and N-free carbon QDs. Insets are the images of carbon QDs aqueous solution in UV lamp irradiation (right) and daylight (left)). (b) PL spectrum of carbon QDs with excitation of various wavelengths documented from 300 to 500 nm [*Source*: https://link.springer.com/article/10.1186/s11671-016-1231-1].

The full scan spectrum has exposed the presence of nitrogen (N 1 s, 400.5 eV), carbon (C 1 s, 284.5 eV), and oxygen (O 1 s, 531.5 eV). Carbon's 1-s has peaked at 284.6, 285.7, and 288.1 eV in the expanded XPS spectrum; they can be allocated to C–C, C–N/C–O, and C=O bonds, in that order. The Figure 4.10 demonstrates peaks of nitrogen 1 – s at 400.3 and 401.6 eV is signifying that nitrogen is present in the form of C–N as well as N–H bonds. The peak of oxygen 1 – s displays 2 peaks at 531.7 eV and 532.8 eV, which are accredited to C=O and C–OH/C–O–C, in that order. It can be concluded from these observations that nitrogen-doped CDs have several different polar groups such as alkyl, hydroxyl, and carboxyl (Ruiz-Rosas et al., 2010).

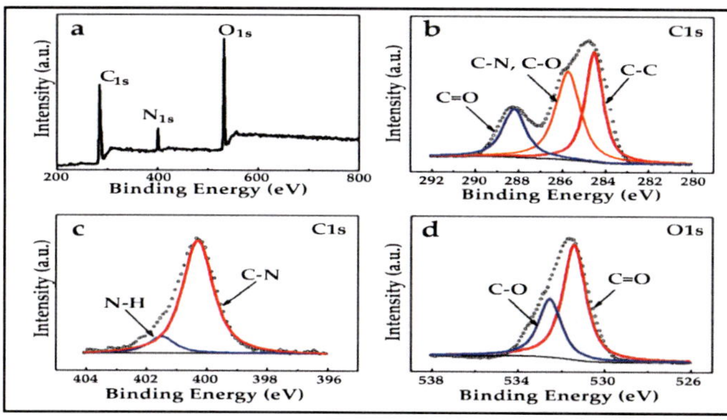

Figure 4.10. XPS survey scan of carbon quantum dots deposited on indium foil. (a) Full scan XPS spectrum. (b) High-resolution XPS survey scan of C1s, (c) High-resolution XPS survey scan of N1s, and (d) O1s region [*Source*: https://link.springer.com/article/10.1186/s11671-016-1231-1].

Figure 4.11 shows the characterization of the morphology of Nano Carbon QDs through AFM and TEM. The TEM images exhibit that Nano Carbon QDs' sizes are largely distributed within the range of 7–15 nm with a typical size of 10.8 nm. The AFM image represents that the topographic height of the Nano Carbon QDs lies in the range of 1–3 nm having major distribution at 2 nm. It further implies that the Nano Carbon QDs have a structure resembling that of the cylinder (like tiny pieces of graphitic layer stacking). For analyzing the relative intensity ratio, a scientist has used Raman spectroscopy around 1.07 of the D-band (1342 cm^{-1}) and G-band (at 1531 cm^{-1}). The resultantly improved visible-light absorption implies that Nano Carbon QDs could be a smart interface material incorporated with porous $TiO2$ for QDSCs (Titirici et al., 2008).

For demonstration, NCQDs unification with anatase (TiO_2) has been investigated by various researchers. Figure 4.12 shows UV–visible diffuse reflectance spectrum of TiO_2 (inset: TiO_2 photo-anode / Nano Carbon QDs). It can be seen that Nano Carbon QD/TiO_2 photoanode get much more improved absorption in visible region which ranges from 400 nm to 550 nm as compared to the pure TiO_2 photo-anode. It must be used for application in r photo-sensitizers. Also, the PL spectrum of TiO_2 and Nano Carbon QDs were also investigated to verify the phenomena (Bhunia et al., 2013; Yan et al., 2015).

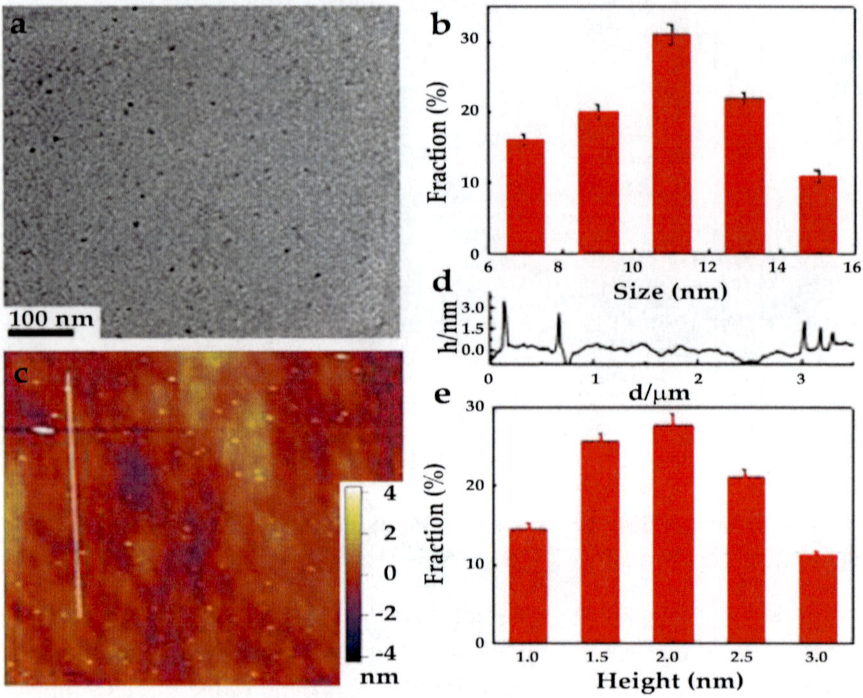

Figure 4.11. (a) TEM image of carbon QDs. (b) Particle size distribution of Carbon QDs. (c) AFM image of Carbon QDs deposited on a silicon substrate. (d) Elevation profile along the white line shown in (c). (e) Height distribution of carbon QDs [*Source*: https://link.springer.com/article/10.1186/s11671-016-1231-1].

Pure TiO_2 has insignificant PL emission under excitation of visible light at 420 nm, whereas, Nano Carbon QDs display a wide and strong emission peak centered at 550 nm. Nonetheless, it is exciting to observe that the emission of Nano Carbon QDs/TiO_2 are quiet weaken in this scenario which implies that TiO_2 hinders the recombination of photo-generated holes and electrons in excited Nano Carbon QDs and absorbs the photo-generated electrons.

Figure 4.12b shows the schematic diagram of the photo-induced electrons transition between TiO_2 and Nano Carbon QDs. The N-dopant of Nano Carbon QDs, in comparison with N-free CDs, can bring in an extra level of energy between π* of oxygen and π of carbon. After excitation, Nano Carbon QDs can absorb the photons having fairly low-energy at a visible range of light and produce the photo-excited electrons. Consequently, there are more electrons which get a transfer to the conducting band of the TiO_2 which leads to the higher current density. For exploring the prospective application in photovoltaic, scientists have fabricated Nano Carbon QDs to be used as sensitizers for QDSCs.

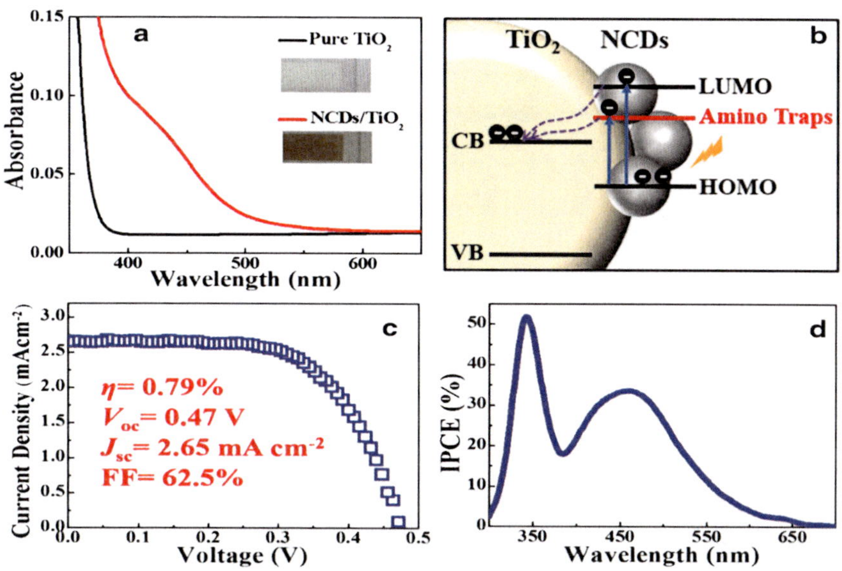

Figure 4.12. UV–visible DRS spectra of Titania (TiO_2) and TiO_2/carbon QD photoanode. Insets are the images of TiO_2/carbon QD and TiO_2 photoanode. (b) A suggested model of photoinduced transition of an electron between carbon QDs (NCDs) and TiO_2. (c) Voltage–Current density characteristics. (d) IPCE action spectra of NCDs-sensitized photovoltaic cells [*Source*: https://link.springer.com/article/10.1186/s11671-016-1231-1].

Figure 4.12c displays the current density-voltage (designated as J-V) curve of Nano Carbon QDs sensitized solar cells under one full sun illumination at AM 1.5 G. The figure shows that the solar cell has a V_{oc} of 0.47 V, J_{sc} of 2.65 mA cm^{-2}, an FF of 62.5%, and overall PCE of 0.79%. Figure 1.5d represents the plot of the conversion efficiency of Incident photon-to-electron (IPCE). It has been observed that the, on the whole, photocurrent response matches strongly with the corresponding absorption spectrum of Nano Carbon QD-sensitized TiO_2 film represented in Figure 4.12a. In between 400 and 550 nm, the value of IPCE is over 10%. It is to be noted that the maximum value of IPCE even reaches 34% in case of Nano Carbon QDs (Yang et al., 2012).

To summarize this chapter, we can state that scientist have developed "green" quantum dot solar cells that are based on nitrogen-doped carbon dots. As demonstrated by experiments conducted by researchers, the finest of such Nano Carbon QD-based solar cells have attained a PCE of 0.79% having V_{oc} of 0.47 V, J_{sc} of 2.65 mA cm^{-2}, and FF of 62.5%. The acquired PCE is the maximum value achieved so far in QDSCs based on CDs. However, studies demonstrate that the more optimization of nitrogen dopant can achieve an improved conversion efficiency of Nano Carbon QD-based solar cells.

CHAPTER 5

MULTIPLE EXCITON GENERATION IN QUANTUM DOTS SOLAR CELLS

CONTENTS

5.1. BACKGROUND

The process in quantum confined semiconductors by which multiple bound charge-carrier pairs are created after absorption of a single high-energy photon is known as multiple exciton generations (MEG). Such effects of charge-carrier multiplication have attained significant attention as they are particularly useful for solar cells in which they can enhance the photocurrent considerably. Current research efforts have verified that more than a single pair of charge carriers per incident solar photon can be extorted out in photovoltaic devices that incorporate quantum-confined semiconductors. Although these proof-of-concept applications highlight the ability of MEG in solar cells, the effect of the carrier multiplication impact on the performance of device remains rather low. This chapter gives a brief review of the latest advancements in the understanding as well as applications of MEG as a photocurrent-increasing mechanism in quantum-dot-based photovoltaics (Ohno et al., 2013; Mayer et al., 2015).

5.2. INTRODUCTION

Large-scale and optimized manufacturing processes for the development of photovoltaic devices (PVs) have promoted technologies in a solar cell to an extent that they are becoming the cheapest form of energy. While materials based on PVs like silicon continue to exceed commercial milestones owing to reduced costs of fabrication and improved market penetration, meanwhile, the power conversion efficiency (PCE) of such devices is primarily limited to almost 33% (Frankl et al., 2010). By incorporating ways to overcome this limit, together with advancements in manufacturing processes of devices, we can develop solar energy technologies to meet the ever-increasing energy demands of the future.

The baseline of basic principles of PVs lies upon the conception of a reversible Carnot heat engine. Thermodynamics itself states the highest efficiency limit of 87% for the conversion mechanism of solar energy into electrical current (Shockley & Queisser, 1961). This value of theoretical performance limit can only be realized in a device architecture where a huge number of sub-cells (each sub-cell optimized for a small part of the solar spectrum) are to be joint in a PV stack and illuminated under the intense concentration of light (Li et al., 2006). In fact, most of the PV devices use only a single cell using one absorbing semiconductor which is responsible for collecting sunlight under non-concentrated conditions. This not o ideal environment for solar cell introduces two dominant mechanisms of loss: first, non-absorption of photons having energy lesser than that of the semiconductor band gap and secondly, speedy cooling of carriers having energy greater than that of the band gap where a fraction of the initial photon energy is lost due to heat (De Vos, 1980). As a result, the ultimate performance of the device is reduced significantly as compared to that of an ideal Carnot heat engine (Sandberg et al., 2012; Stewart et al., 2013).

Two researchers, Shockley and Queisser (1961) have carried out significant work in which they have quantified the obvious loss processes in a comprehensive balance model. Their research expresses the resulting impact on the maximum achievable limit of efficiency of the solar cell in their band gap-reliant Shockley-Queisser model (Ten Cate et al., 2015; Midgett et al., 2010; Gao et al., 2012).

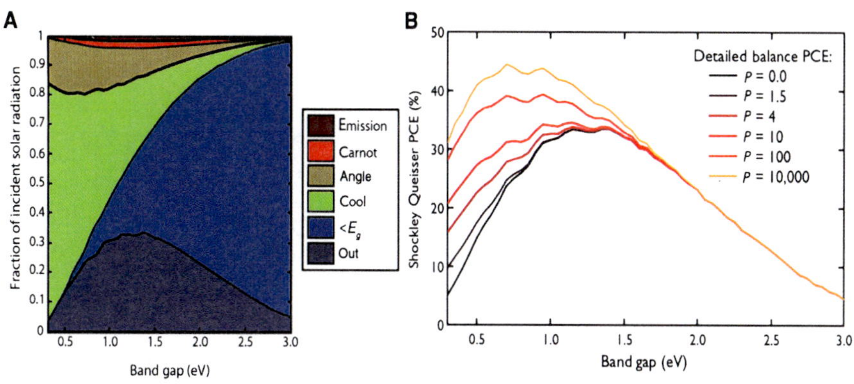

Figure 5.1. Shockley-Queisser limit: loss methods and enhancement in potential efficiency by MEG. (A) Breakdown of the various loss processes resulting in the band gap-dependent Shockley-Queisser limit for single junction solar cells (out, dark blue). <Eg (light blue) and cool (denoted by green) displays the losses in energy because of incomplete absorption and fast cooling of hot carriers, respectively, angle (denoted by orange) and emission (denoted by dark red) are entropic losses related to angle restriction and incomplete trapping of light, respectively, and Carnot (denoted by light red) shows the thermodynamic efficiency limit. (B) Theoretical improvement in the limit of the Shockley-Queisser for solo junction solar cells owing to MEG where the parameter P tells the rates for MEG (kMEG) as well as hot carrier thermalization ktherm by the expression P=kMEG/ktherm. (Polman & Atwater, 2012) [*Source*: https://www.degruyter.com/view/j/nanoph.2018.7.issue-1/nanoph-2017-0034/nanoph-2017-0034.xml].

As light absorbs in the active layer of a solar cell, it generates excited bound pairs of charge-carrier (i.e., excitons). Before we can extract these through an external circuit, they repeatedly encounter loss processes that decrease the efficiency of the solar cell. Angle restrictions or incomplete trapping of light is responsible for up to 20% of all losses of energy in the SQ limit. However, it is, mostly rapid carrier cooling and incomplete absorption which is responsible for setting the SQ limit for single junction solar cells to the highest PCE of 33.7% for a semiconductor having $E_g \approx 1.3$ eV (Figure 5.1).

Among the most capable concepts to repress loss processes connected to incomplete absorption are mechanisms of photon up-conversion where recovery of low energy photons below the band gap is carried out for the operation of solar cells (Beard et al., 2012; Mahato et al., 2015). However, as this concept does not lie within the main focus of this chapter, we will refer the interested reader to other reviews that address this topic in detail (Klimov et al., 2000; Murphy et al., 2006; Wu et al., 2016).

The most extensively used approach for decreasing losses from fast cooling is through multi-junction PV devices. Such devices make use of multiple (generally 2 or three) semiconductors having different band gaps to best match with the solar spectrum (Baur et al., 2007; Schulze & Schmidt, 2015). Multi-junction cells have created efficiencies of PV lying far beyond the SQ limit for mono junction solar cells (so far 42% under un-concentrated AM 1.5 G conditions). However, their high cost of fabrication still makes large-scale deployment challenging (Dimroth et al., 2014). Another concept employed for minimizing losses owing to hot carrier cooling is the fast extraction of the initial hot carriers. Even though this phenomenon has been speculated to fabricate superior photo-voltages, still, experimental evidence for the effect's advantages in an environment of a solar cell has not yet been produced (Antončík, 1957; Nozik, 2002; Schaller & Klimov, 2004).

Another concept which is capable of reducing carrier-cooling losses is Carrier multiplication (CM) effects; in this method, the energy of a high-energy photon is disseminated among multiple pairs of carriers having lower energy (Vavilov, 1959; Kolodinski et al., 1993). In such case, additional charges are formed from the energy that else would have been lost in the form of heat. This kind of processes have been demonstrated to take place proficiently in inorganic quantum dots (QDs) and organic semiconductors, where the phenomena have been termed as multiple exciton generations (MEG) and singlet exciton fission, respectively (Merrifield et al., 1969;

Brendelk et al., 1996; Wilson et al., 2013). According to calculation, such CM mechanisms can increase the maximum efficiency of solo-junction PVs from around 33% to approximately 44% (see Figure 5.1) (Smith & Michl, 2013).

Various materials and combinations of materials have been proposed to utilize CM effects in solar cells. Singlet exciton fission materials together with appropriate QD-based solar cells and charge acceptors have both been shown to generate external quantum efficiencies (EQEs) above 100% in a PV device (Hanna & Nozik, 2006; Congreve et al., 2013; Tabachnyk et al., 2013). Despite the fact that such proof-of-concept studies demonstrate that CM effects can enhance the photocurrent of a solar cell, still, PCEs in both types of the device remains small (Yang et al., 2014; Semonin et al., 2011; Davis et al., 2015). Particularly in case of MEG, the latest spectroscopic observations propose a combination of photon energy-dependent MEG efficiency and high MEG threshold energies (i.e., MEG efficiency increases in a slow manner after the threshold energy instead of increasing as a step function) as the prime reasons for the unpretentious contribution to PV power conversion (Smith & Binks, 2013; Bo̍hm et al., 2015).

Several comprehensive review articles in the literature summarize the present understanding of the MEG mechanism itself (Beard et al., 2010; Padilha et al., 2013). In this chapter, we will just briefly talk about mechanistic details and will mainly focus on current progress towards a proficient integration of MEG in operational solar cells. Following to a concise introduction of the process in their nano-scale environment, we will further discuss the underlying physics of devices of lately reported QD-based solar cells that exhibit the involvement from MEG to the photocurrent. In the end, this chapter provides some device and material-related suggestions to improve the impact of MEG in future PVs which might enable performance of device beyond the SQ limit (Robbins, 1980; Doany & Grischkowsky, 1988; Price et al., 2015).

5.3. QD MATERIALS FOR MEG IN PV DEVICES

In addition to quantum confinement effects, there are several other properties of the material which should be considered for MEG on the device as well as process level. Ideal materials will demonstrate comparatively slow cooling of the hot carrier, to enlarge the time window for MEG (Brown et al., 2014; Chuang et al., 2014; Liu et al., 2017). Moreover, by adjusting the Shockley-Queisser calculations for incorporating MEG, two researchers namely,

Nozik and Hanna have demonstrated that the ideal band gap of photovoltaic material is decreased from approximately 1.3 to 0.8 eV. Materials having relatively small bulk band gaps are required for providing excellent tunability of the quantum confinement in this spectral area (Bude & Hess, 1992; Wolf et al., 1998).

Even though MEG has been detected in various materials, such as silicon, most of the studies focus on QDs that consist of lead chalcogenides (PbS, PbSe, and PbTe) (Shabaev et al., 2006; Cunningham et al., 2011). Lead chalcogenides have large Bohr radius as compared to, for example, GaAs or silicon; this large radius permits strong quantum confinement effects as well as efficient tuning of the confinement for producing a band gap in the near-IR (infrared) region (0.7–1.0 eV) (McGuire et al., 2008; Padilha et al., 2013). Additionally, suitable ligand exchange methods have been developed for these particles, providing control over the properties of the QD superstructure (for example, energy alignment, inter-particle spacing, etc.) (Alivisatos, 1996; Bartnik et al., 2010). In combination with slow cooling of the hot carrier, lead chalcogenide QDs is a precious material platform to investigate parameters related to MEG, not just on the process itself, but on the PV device level as well (Christensen, 1976; Würfel, 1997; Nozik, 2001).

5.4. MEG IN QD-BASED SOLAR CELLS

Having overviewed the most relevant theoretical aspects that might enhance the photocurrent of QD-based solar cells through MEG, now, we will focus on practical exhibition (demonstration) of these concepts in operational PV devices. So far, three reports in the literature show the extraction of more than single charge carrier per incident photon in PVs based on QD (Ellingson et al., 2005; Rupasov & Klimov, 2007). Hence, such devices provide proof-of-concept studies highlighting that MEG can enhance the photocurrent of operational solar cells for sure. The target of each of these reports is a different concept for modulation of the carrier kinetics within the QD film to improve the effect of MEG in a working device (Ellingson et al., 2005; Böhm et al., 2014; 2015). Additionally, as all three reports make use of different materials and different material structures, they follow the traditional PV device concept of a worn-out hetero-junction. Since the latter architectural considerations of traditional QD-based solar cells do not lie in the focus of this chapter, we will like to refer the interested reader to other articles available in the literature (Tang & Sargent, 2011; Ning et al., 2015).

5.4.1. Tailoring the QD Ligand Chemistry

In his report, Semonin et al. (2011) take into consideration such PV devices where PbSe QDs having small, di-thiol ligands molecules are paired with a MO that accepts an electron. As stated in the report, QD layers nearest to the anode were treated by ultra-short diamines (see Figure 5.2A). For QDs having suitably narrow band gaps, he measured peak EQEs beyond 100% (Figure 5.2B). As stated earlier, application of either di-amine or di-thiol as the single QD ligand does not produce such high values of EQE (Robel et al., 2009; Beard, 2011).

Taking into consideration the high absorption cross-section of QDs in the energy spectrum relevant to MEG, ($h\nu_{photon} > 2E_g$), previously it has been argued that photons which contribute to MEG are just absorbed in an extremely thin layer, near to the electron- accepting MO. Experiments reveal that such a proximity to the interface shows favor to ultra-fast charge injection (nearly <50 fs) which hinders MEG processes of loss such as Auger recombination by extricating charge carriers before they can recombine (Kramer & Sargent, 2013; Carey et al., 2015). Notably, tailoring the energetics between QD donor and MO acceptor using surface ligands considerably affects this rate of electron transfer (Leatherdale et al., 2002; Yang et al., 2011). Normally, charge transfer rates are not just subjective to the donor-acceptor energetic, but also get influenced by the coupling strength among the orbitals that are involved in the transfer process itself; in that way permitting faster transfer of charge in stronger coupled partners. Moreover, application of hydrazine as ligands for particles near the hole-extracting electrode results in shifting of the QD energetic nearer to the vacuum level (see Figure 5.2A). It follows that the depletion region present within the QD film is extended, which enhances the efficiency of charge extraction.

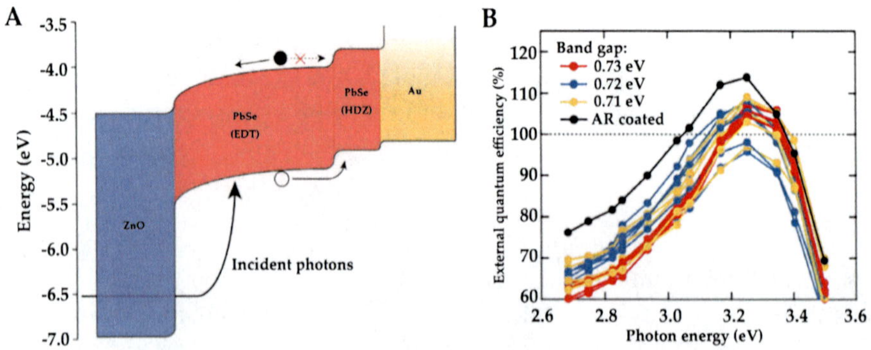

Figure 5.2. The architecture of device and EQE values >100% for solar cells containing PbSe quantum dots. Energy landscape of a solar cell developed from PbSe QDs that have been treated by 1,2-ethanedithiol (EDT) or hydrazine (HDZ). The variation in doping density in QD film and an electron acceptor (ZnO) creates a depletion region inside the QD layer, which impels charge extraction after the generation of free carriers. (B) External quantum efficiency (EQE) of the same PV device architecture exhibiting the high-energy region for three different sizes of PbSe QD. All sizes of QD show the generation of more than single carrier produced for each incident photon at $h\nu_{photon}$ = 3.3 eV. [*Source*: https://www.degruyter.com/view/j/nanoph.2018.7.issue-1/nanoph-2017-0034/nanoph-2017-0034.xml].

Another effect of employing hydrazine as a ligand is the exceptionally short interparticle spacing; this decreases the width of the tunneling barrier among QDs and in doing so increases the carrier mobility of the entire film. Both these effects are mainly used for holes that are created through MEG as they are usually located far-off from the hole-extracting electrode and are, therefore, likely to recombine during charge extraction.

Regardless of the beneficial effects of a mixed ligand strategy to encourage hole-extraction, it is also responsible for reducing the attainable photo-voltage of the device. In a traditional QD-based solar cell, the voltage is primarily defined via the level of quasi-Fermi splitting between the (n-type) MO acceptor and the (mostly p-type) QD donor at the p-n junction. The enhanced n-type doping density of the QD film brought in by the hydrazine ligands decreases this splitting and thus lowers the extractable photo-voltage. This is mostly problematic since the V_{oc} in QD-based solar cells is already comparatively poor owing to other effects like interfacial trap-states and sub-bandgap QD trap states (Anderson & Lian, 2005). Moreover, as ultra-short ligand molecules improve charge extraction, meanwhile, they also deteriorate the quantum confinement within the particles, which eventually interferes with the MEG mechanism itself.

Hence, at this stage, we can conclude that the stated PbSe QD-based devices in the literature provide a first proof-of-concept study for the ability of MEG to increase the photocurrent in solar cells. Nonetheless, the depleted hetero-junction device architecture represents a concern in that context: as strong quantum confinement is needed for efficient MEG, at the same time, this confinement ought to be weak enough to permit ample inter-particle coupling for efficient charge extraction, particularly a hole-extraction.

5.4.2. Tuning the Shape of the QD

Quantum-confined, one-dimensional nanostructures (that is, nano-rods) exhibit improved efficiencies of MEG over spherical QDs not just on the level of a single particle, but in devices as well. Mainly, this has been accredited to two effects: first being the stronger Coulomb matrix element in nano-rods, which increase MEG without changing the reverse process (Auger recombination, to be precise). (Morgenstern et al., 2016). Secondly, the length of the nano-rods themselves helps charge transport as lesser inter-particle hopping steps are needed to reach the electrodes (Chuang et al., 2015).

In addition to the challenge of adequate surface passivation in these nanostructures, it seems that the development of enough quantities of high-quality samples of nano-rods from materials pertinent to PV applications is not easy. Along with other reasons, the centrosymmetric crystal structure of most extensively employed low-bandgap QD materials (for example, the lead chalcogenides PbSe, PbS, or PbTe) restricts the yield of pure nano-rods throughout synthesis and promotes the development of side products like nano-hooks and nano-crosses that are hard to separate (Aerts et al., 2013).

In spite of these synthetic hardships, the latest report demonstrating devices comprising of high-quality PbSe nano-rods verifies that MEG in solar cells comprising of non-spherical nanostructures affects the photocurrent more strongly comparative to spherical, analogous Q D-containing samples. In these devices, it can be particularly noted that only the solar cells where the nano-rods are treated with organic ligands and halides show clear proof for MEG in the EQE spectra which again stresses the significance of cautious surface passivation. Moreover, comparison of the internal quantum efficiency (IQE) or absorbance-corrected QY of devices containing nanorods exhibits significantly enhanced MEG performance over spherical QDs (see Figure 5.3). Amusingly, in the solid state, PbSe nano-rods demonstrate a sharper rise in QY with photon energy comparative to corresponding nano-rods in solution even with the fact that IQEs of the device are normally abridged by Auger recombination contending with charge separation and by mechanical device recombination losses (see Figure 5.3). Similar observations are reported for sphere-shaped QDs where the MEG yield of particles in solution as well as the film has been observed through microwave conductivity measurements (Huynh et al., 2002). Likely explanations for this incident range from the creation of inter-particle QD band structures in the solid state to possible trap-assisted MEG mechanisms (Boercker et al., 2013).

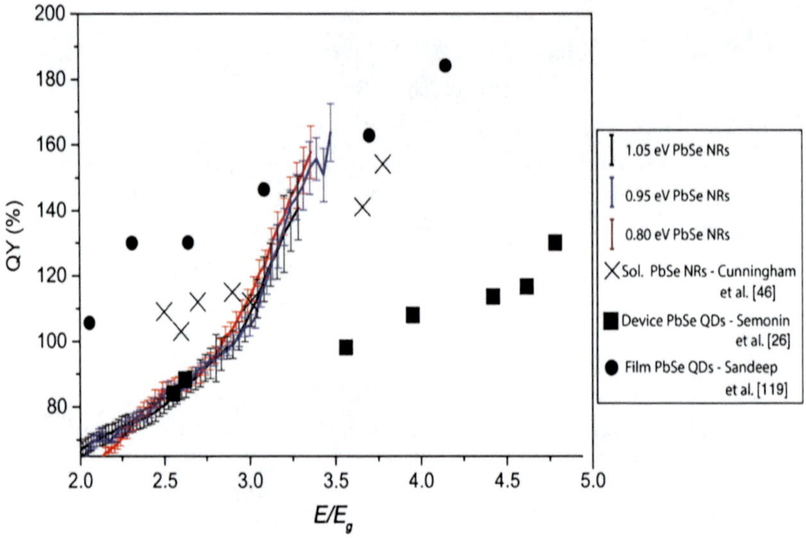

Figure 5.3. Internal quantum efficiencies (IQEs) of devices created from PbSe nanorods of three different band gaps (indicated by solid colored lines) and peak quantum yields of devices comprising of spherical PbSe QD (indicated by squares). The MEG quantum yields of nano-rods (based on PbSe) in solution and of sphere-shaped PbSe QDs in the solid state are indicated as crosses and circles, respectively. [*Source*: https://www.degruyter.com/view/j/nanoph.2018.7.issue-1/nanoph-2017-0034/nanoph-2017-0034.xml].

One more complexity in using nano-rods as the major absorber in solar cells includes the deposition of films with sufficient optical density and enough vertical carrier mobility. In films of nonzero dimensional QDs (such as tetrapods or nano-rods), experiments show that local "dead ends" in arbitrarily deposited nanostructures enhance the probability for carrier recombination hence limiting carrier extraction (Sandeep et al., 2013). Nevertheless, large-scale vertical ordering through self-assembly techniques has been established for high-quality samples of nanorod (Allan & Delerue, 2009). If we can fabricate such a nanostructured film architecture by using first chalcogenide nano-rods, it will permit efficient vertical charge extraction at the same time as taking benefit from the high MEG yields in nanorods.

In addition to nanorods and QDs,a recent evidence for MEG taking place in two-dimensional nano-sheets is reported (Hindson et al., 2011). Experiments demonstrate that these nanostructures efficiently convert the photon energy more than E_{th}^{MEG} into surplus charge carriers, however, these two- dimensional nano-sheets need very high photon energies to release MEG ($E_{th}^{MEG} > 3E_g$) (an effect, which leaves two-dimensional nano-sheets unproductive as sole photon absorbers at this stage.

5.4.3. The Selection of QD Material

At initial times, the efficiency of generating multiple exciton states is chiefly influenced by a kinetic competition amid the MEG process and super-fast cooling of the hot carrier of the photo-generated state (Baker et al., 2009; Aerts et al., 2014). At later stages, the multiple exciton states may either recombine (for instance, through an Auger process) or separate into free carriers. In case of lead chalcogenide QDs, the latter process proceeds on time scales of 20–200 ps (Schaller et al., 2005). As stated earlier, if an ionizing interface lies near the generated exciton, the charge separation can be no less than two orders of magnitude faster (Allan & Delerue, 2006; 2009). Taking into consideration, the high absorption section of QDs in the spectral region significant for MEG (i.e., $h\nu_{photon} > 2E_g$), the majority of multiple-exciton states are easily generated near an interface where a charge accepting material can compel efficient separation. It follows that the multiple-exciton states' separation efficiency is not methodically limiting additional improvements from MEG that contributes to the photocurrent (Nootz et al., 2011; Miller et al., 2012).

We can draw a different conclusion for the early time yield of MEG, where different materials strongly influence the generation of multi-exciton states through MEG. For example, spectroscopic results after analyzing MEG yield in lead chalcogenide QDs for few picoseconds indicates a sequentially rising CM yield within the series PbSe, PbS, and PbTe. Even though in experiments, these QY measurements have been conducted on particles having an alike band gap, the altering Bohr radii of the materials implies that this does not equate to the same extent of confinement such that the extent of quantum confinement raises from PbS to PbTe. Hence, the experimented rise in QY follows the growing Bohr exciton radii for PbSe (46 nm), PbS (18 nm), and PbTe (150 nm). Moreover, the latest reports substantiates that this tendency of enhanced MEG yield is also existent in solar cells: incorporating PbTe QDs into operational PV devices shows peak IQEs and EQEs (155±7% and 122±4%, respectively), considerably higher as compared to devices which consists of PbSe QDs of similar QD band gap (ca. 95% and 80% respectively) (Müller-Steinhagen & Trieb, 2004; Lan et al., 2016; Liu et al., 2016).

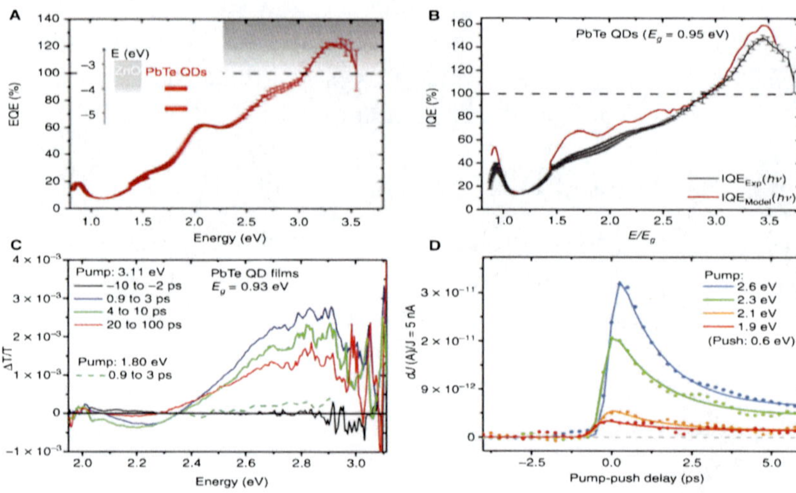

Figure 5.4. Proof of carrier multiplication in solar cells containing PbTe quantum dot. (A) External and (B) internal quantum efficiencies of solar cells based on PbTe QD. The architecture of the device is sketched in the inset of (A). IQE-Model (hv) and IQE Exp (hv) are the internal quantum efficiencies determined by transfer matrix modeling and by reflectance measurements, respectively. (C) Conduction of pump-probe transient absorption experiment on films of PbTe QDs signifying a transiently stable carrier population at around 2.8 eV. (D) The response of photocurrent as a function of time delay in between excitation pulse and succeeding low-energy push pulse, which re-energizes the new carrier population. [*Source*: https://www.degruyter.com/view/j/nanoph.2018.7.issue-1/nanoph-2017-0034/nanoph-2017-0034.xml].

There is an argument based on the stronger quantum confinement existent in heavier lead chalcogenide QDs that, the stronger matrix element forcing the energy transfer process from single exciton state to multiple-excitonic states can clarify the higher QYs for PbTe QDs (Wang et al., 2013; Thompson et al., 2014). For the reverse procedure (i.e., Auger recombination), which is ruled by the same matrix element, no variation in timescales of recombination as a function of evident quantum confinement is being observed in the series PbSe, PbS, and PbTe (Jean et al., 2013; Tamang et al., 2016). Taking into consideration, the ultra-fast exciton separation near QD-MO interfaces and material independent Auger kinetics, it is believable that high MEG yields for weighty lead chalcogenide QD devices are because of the stronger quantum confinement permitting, the faster buildup of multiple-excitonic states. After that, these states are efficiently separated before occurrence of loss processes like Auger recombination (Sun et al., 2009; Tabachnyk et al., 2014; 2016).

Another speculation for explaining the yields of material-dependent MEG in lead chalcogenide QDs is the declining frequency of longitudinal optical (abbreviated as LO) phonons in the series PbSe, PbS, and PbTe (Dalyan et al., 1971. If the frequency of LO phonon turns out to be too small to bridge the gap of energy in between the discrete and energetically alienated excited states, we then require multi-phonon processes to cool the hot carriers which results in reduced cooling rates (Bode & Levinstein, 1954; Beard et al., 2007; Law et al., 2009). As stated earlier, such arguments might clarify slow carrier cooling at the edge of band where the quantum confinement certainly produces discrete and separated states (Jaeger et al., 2012; Trinh et al., 2012; Damtie et al., 2016). These arguments do not hold for the MEG relevant energy regime (that is energetically detached from the discredited near-band-edge states), and thus are unable to explain the noticeable difference between the quantum efficiencies of PbTe and PbSe QD-based solar cells. Still, current spectroscopic measurement on lead chalcogenide QDs is consistent with a briefly stable carrier population far-off from the band edge in the vigorously condensed energy manifold (Midgett et al., 2010; Trinh et al., 2011; Liu et al., 2013).

The contribution of the MEG towards enhancing the photo-current of a photovoltaic device is advantageous only if it is present above E_{th}^{MEG} of the cell. Efficient transient photocurrent spectroscopic approaches followed on operational PbTe QD photovoltaic cells along with the IQE characteristics of the device suggest that PbTe QDs display the location of E_{th}^{MEG} at comparable energy values to the bottleneck (Figure 5.5).

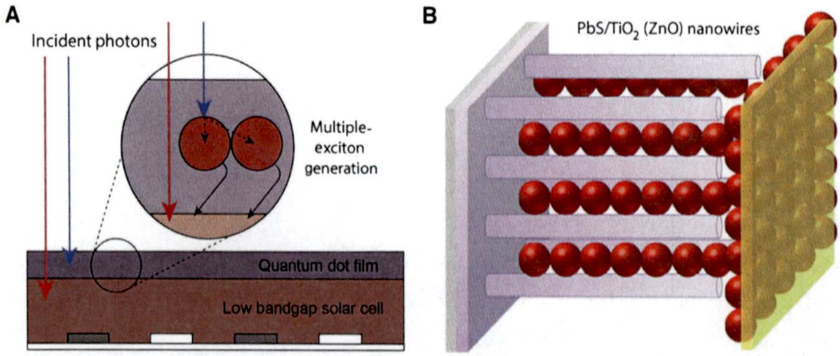

Figure 5.5. Potential methods for upgrading state-of-the-art solar cells utilizing carrier multiplication in quantum dots. Schematics of (A) a MEG-enhanced, interdigitated bottom-contact solar cell. High-energy photons (blue arrows) are absorbed in the thin QD layer processed on top of a bulk semiconductor where they can undergo MEG. The carriers from the generated multi-exciton state are subsequently transferred to the low band gap semiconductor and can contribute to the photocurrent. Meanwhile, photons of lower energy are more likely to pass through the QD film due to the lower absorption cross-section and can be absorbed and converted into charge carriers in the bulk semiconductor. (B) A device concept based upon a nanostructured hetero-interface between QD absorber and metal oxide acceptor. Here most multiple-exciton states are generated close to an ionizing interface. Both an efficient harvest of multi-exciton states and long optical path lengths for absorbing solar photons are possible. [*Source*: https://www.degruyter.com/view/j/nanoph.2018.7.issue-1/nanoph-2017-0034/nanoph-2017-0034.xml]

CHAPTER 6

GRAPHENE AND QUANTUM DOTS COMPOSITE FOR PHOTOVOLTAIC DEVICES

CONTENTS

6.1. BACKGROUND

Scientists have carried out fabrication of wide-area flexible and translucent optoelectronic devices through non-covalent interaction in between single-layer graphene and CdSe/ZnS core/shell quantum dots. The photo-induced transfer of charge has been extensively verified by photoconductivity measurements and fluorescence quenching experiments (Novoselov et al., 2004; 2005; Neto et al., 2009). Quantum dots having been decorated with graphene device have shown outstanding photo-switching property which shows their potential to be used as a long wavelength sensitive photodetector. Moreover, scientists have observed that 1-D (graphene)/0-D (QD) hybrid structure helps in faster transfer of electrons. Experiments have shown that when this hybrid material is used as photo-electrochemical cell anode, it exhibits both fast photo-switch response and high photon to current conversion efficiency (IPCE).

6.2. INTRODUCTION

One of the promising devices for future energy applications are Photovoltaic devices. For the development of photovoltaic devices, efficient dissociation of hole and electron is critical. One of the major challenging obstacles in developing such devices is the recombination of electron-hole pairs. Decomposition of photo-generated holes or electrons is mostly caused by radiative, i.e., luminescence, as well as non-radiative recombination, i.e., defects, grain boundaries, or other sites related to weaker chemical bonding. Hence, there are various critical factors that must be taken care of when selecting suitable materials for separation of photogenerated electron-hole pairs having high efficiency:

- After formation of exciton, high crystallinity for avoiding recombination in crystals;
- Suitable work function of conductor adjustable to band-gap of semiconductor;
- Fast transfer of charge from semiconductor to conductor;
- High carrier concentration or high conductivity or for the conductor.

Graphene was discovered in 2004, since then, it has been used in many different areas owing to its promising properties, such as high mobility, large interface, and tunable band gap (Geim & Novoselov, 2007; Han et al., 2007;

Li et al., 2008). All of these properties allow graphene to be a very capable electron acceptor in photovoltaic devices. One of the significant applications of graphene is its use as a photodetection device which may perhaps pave the way for developing solar cells, LEDs, and photo-switches, etc. (Shi et al., 2009; Lin et al., 2011) (Figure 6.1).

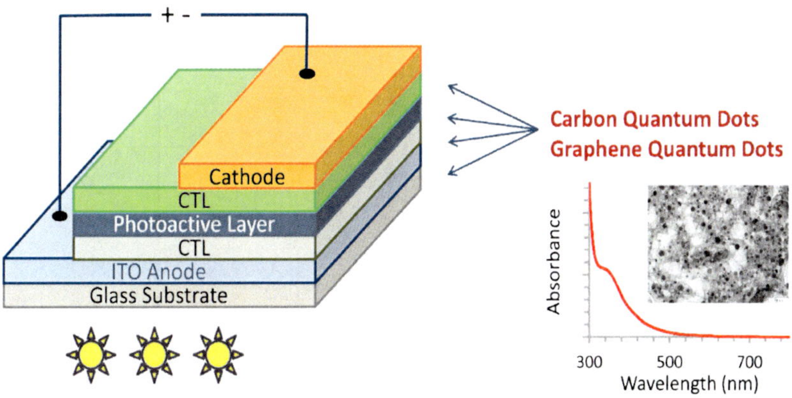

Figure 6.1. Graphene and Carbon QDs based photovoltaic cell [*Source*: http://www.mdpi.com/2079-4991/6/9/157].

Quantum dots (QDs) have been extensively studied for photovoltaic devices due to their size-tunable optical properties (Robel et al., 2006; 2007; Kongkanand et al., 2007). Furthermore, their capability to produce multiple electron-hole pairs through inverse Auger processes and attain higher conversion efficiency energy above the Shockley–Queisser limit make easy to use them as a sensitizer in photovoltaic conversion or as light harvesting assembly (Sun & Greenham, 2006; Kamat, 2008; Rühle et al., 2010). Numerous methods have used for attaching QDs on the surface of carbon materials (Sun et al., 2005; Gur et al., 2007; Gap, 2010; Peng, 2010). The conductivity of carbon materials cab ne kept through non-covalent bonding attachment which makes it capable for use in device applications (Regulacio & Han, 2010; Hungría et al., 2008; Juarez et al., 2008).

Hybrid structures such as CNT/QD have been extensively studied owing to their prospective photovoltaic applications (Iacono et al., 2013; Na et al., 2008; Lu et al., 2009; Li et al., 2010). Improved distribution of QD on the surface of carbon would surely get us better photo-conversion efficiency. Lately, fast and efficient transfer of energy from (CdSe)ZnS to graphene had been reported with a detailed study of fluorescence from individual CdSe/ZnS nanocrystals in touch with single and few-layer sheets of graphene

(Pasricha et al., 2009; Swathi & Sebastian, 2009; Zhou et al., 2009). Moreover, numerous studies have been carried out on reduced graphene oxide (RGO)/QD hybrid material which exhibits the high potential of this research field (Jeng et al., 2010; Wang et al., 2010; 2011). A method known as Chemical vapor deposition (CVD) graphene offers large area graphene reaching up to 30 inches as well as mobility up to 3000 $cm^2/v^{-1}s^{-1}$, which makes it a fine candidate for such photovoltaic devices (Chen et al., 2010; Graphene & Dots, 2010; Gaudreau et al., 2013). Furthermore, the lesser conductivity of RGO may put a stop to more upgrading of photovoltaic devices (Cao et al., 2010; Dai, 2010; Alaferdov et al., 2014). CVD graphene seems capable for many different applications of electronics. For a solar cell, it has been used as a transparent current collector (Moon et al., 2008; Kim et al., 2010; Zedan et al., 2010). Solar cells which incorporate the nanoparticle-graphene hybrid will have following benefits:

- High mobility of electron/hole.
- Greater lifetime of exciton owing to the effective separation of charge from the nanoparticles to the graphene.
- Proficient absorption of light.
- Efficient transfer of charge from particle to the current collector.

In this chapter, pyridine functionalized (CdSe)ZnS QDs mono-layer have been stated which attach to single-layer graphene for various photovoltaic device applications (El-Den & El-Nahass, 2003; Bang et al., 2006; Hezam et al., 2010). Transfer of electron from QDs to the graphene offers us a substitute for the development of photo-device. The fast photoresponse of this FET device may be used as photodetector whereas; the use of high IPCE of the hybrid material as the anode of photo-electrochemical cell (PEC) signifies its high potential for development of solar energy (Darakchieva et al., 2000; El-Nahass & El Den, 2001; Al-Kuhaili, 2004).

6.3. FABRICATION AND CHARACTERIZATION

6.3.1. Synthesis of Graphene

Scientists have synthesized CVD graphene by using a copper foil as the catalyst (Fadel et al., 1998; Kucur et al., 2003; Khoshman & Kordesch, 2006). In this method, normally, the copper foil can be etched with acetic acid for 10 min at 35°C to take away any CuO. After washing the foil thoroughly using deionized (D.I.) water, the samples of copper foil can then be heated in a 2-Torr Ar/H_2 (200:200 sccm) atmosphere up to 1000°C and afterward thermally annealed for 30 minutes (Querner et al., 2005; Hu et al., 2008; Inamdar et al., 2008). It is then exposed to methane (100 sccm) mixed with H_2 flowing under 20 Torr at 1000°C for 20 minutes; afterward, the temperature is cooled down to almost 25°C at a cooling rate of 20°C per minute. Subsequently, the underneath copper foil is then etched in an aqueous solution of FeCl3 (0.5 mol/dm^3) and then thoroughly rinsed using HCl (3%) and D.I. water, respectively. Finally, the graphene film is transferred to a Si substrate that is covered with 300 nm thermally developed oxide for additional characterization (Guldi et al., 2006; Juárez et al., 2007; Li et al., 2009).

6.3.2. Production of Graphene Based Field Effect Transistor Device (GFET)

Many scientists have used oxygen plasma to pattern and etch the CVD graphene into a rectangular shape through photolithography and a Reactive Ion Etch (abbreviated as RIE) system. Two terminal devices can be fabricated with Ti/Au (2 nm/80 nm) electrodes, followed by the evaporation of the device with a Ti/Au (2 nm/100 nm) back gate. In this experiment, the total distance between the drain and source is typically set to be 70 μm; whereas, the width of the graphene layer was set around 50 μm (O'regan & Grätzel, 1991; Grätzel, 2001; Shim et al., 2011).

6.3.3. Characterization Methods

For characterization, researchers have employed Raman spectra by using a Renishaw micro-Raman spectrometer having a 532 nm visible laser as the source of excitation (Fujishima & Honda, 1972; Orlanducci et al., 2006). Atomic force microscopy (AFM) can be conducted using a Multimode-5, Veeco instrument which is operated in the tapping mode using standard silicon cantilevers. UV-visible spectra can then be measured using a Perkin Elmer Lambda 35 UV/VIS Spectrometer. Afterward, Photoluminescence spectra can be recorded using Spex Fluorolog Tau-3 fluorescence spectrophotometer. By employing a BD Pathway 855 HT Confocal microscope, fluorescence imaging can be performed. For this imaging, the sample can be irradiated with the light source coming from an arc lamp and filtered through a 488 nm (±10 nm) bandpass filter (Yang et al., 2007; Liu & Zeng, 2008). The fluorescent light that is released from the sample can then passed via a 495nm dichroic mirror and an emission filter of 624 nm (±40 nm), followed by detection using a CCD camera. For imaging of the whole sample using the 2x objective, a scientist has built a montage of individual images, each of which consisted of 1024 x 1344 pixels and represented a 271 x 355 mm^2 area. Additionally, the 20x objective was employed for high-resolution imaging in which each pixel covers a 312 nm^2 area (Chang et al., 2010; Guo et al., 2010; Tang et al., 2010). The impact of uneven illumination of the individual images can be removed from the final montage image by the division of each image through a background image (image of a homogeneously fluorescent sample gathered using the same optical path) according to the following equation:

$$I_{flat}(x,y) = \frac{I_{original}(x,y)}{I_{background}(x,y)} \times I_{background}$$

6.3.4. Photo-Electrochemical Measurements

Measurements regarding photo-electro-chemical can take place in a three-compartment cell with the help of a potentiostat, using a working electrode, a saturated Ag/AgCl reference electrode, and a Pt wire gauze counter electrode. This specific configuration makes it possible to carry out measurements related to photocurrent under electrochemical bias. For the recording of I-V characteristics, a potentiostat or a galvanostat can be used, whereas, for generation of photocurrent density, a 30W white source of light can be used (Pagona et al., 2008; Mountrichas et al., 2007; 2009). In this kind of experiments, NL/< the intensity of illumination close to the surface of the electrode is taken to be 30 mW cm^{-2}; while the electrolyte is 0.1 mol/ dm^3 aqueous Na_2S.

Under UV lamp illumination (365nm), the calculation of IPCE value is done by normalizing the densities of photocurrent for incident light intensity and energy and according to the following expression:

$$IPCE (\%) = 100 \times 1240 \times i/(W_{in} \times \lambda_{ex})$$

In this equation, W_{in} is the incident light intensity (W/cm^2), i is the photocurrent density (A/cm^2), whereas λ_{ex} is the excitation wavelength (nm).

6.4. INTERPRETATION OF CHARACTERIZATION DATA

6.4.1. Band-Gap Feasibility

Figure 6.2 illustrates the configuration of the device that has been used by a scientist for transistor and r resistance measurements. To achieve most favorable band-gap engineering of the present device, we should select CdSe@ZnS core/shell structure quantum dots having absorption at 630 nm. The photostability of the QDs can be considerably increased by the coverage region of a ZnS shell on to the CdSe cores (Li et al., 2011; Cloke, 2015).

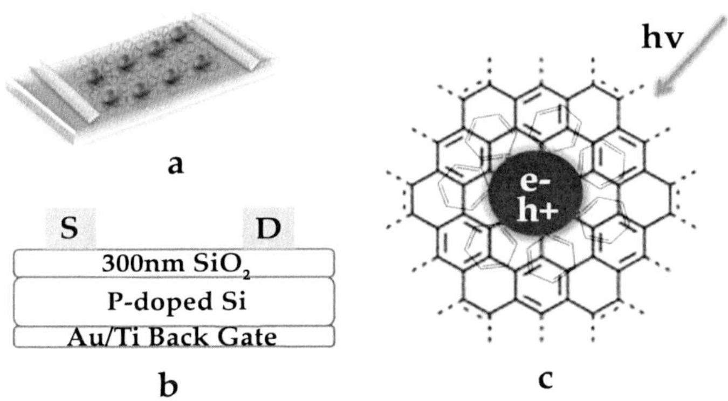

Figure 6.2. Schematic diagram of graphene photoswitch devices. a) Top b) side view of QDGFET devices, c) exciton creation in QD and transfer of an electron to graphene with light illumination. [*Source*:https://www.researchgate.net/publication/275012245_Graphene-Quantum_Dots_Composite_for_ Photovoltaic_Devices].

By using CVD method, scientists have successfully fabricated large area single-layer graphene. As can be seen in Figure 6.3, the appearance of G (1560–1620 cm^{-1}), D (1300–1400 cm^{-1}), and 2D (2600–2800 cm^{-1}) bands verifies the graphene layer. The intensity proportion of 2D/G is nearly 2, which indicates single-layer graphene.

Mixtures of pyridine functionalized (CdSe)ZnS nano-particles can be prepared plainly by addition of ethanol into octylamine capped (CdSe)ZnS solutions; afterward, it is rinsed quite a few times with pyridine through a centrifuge, ultimately followed by sonication for 5 minutes at room temperature. The fabrication of hybrid film hybrid can be simply carried out by dropping as-prepared solution onto a layer of graphene. After that, the film is to be rinsed 3 times to get free from particles not attached to graphene through π-π interactions (Kyle et al., 2011; Ghazinejad et al., 2013).

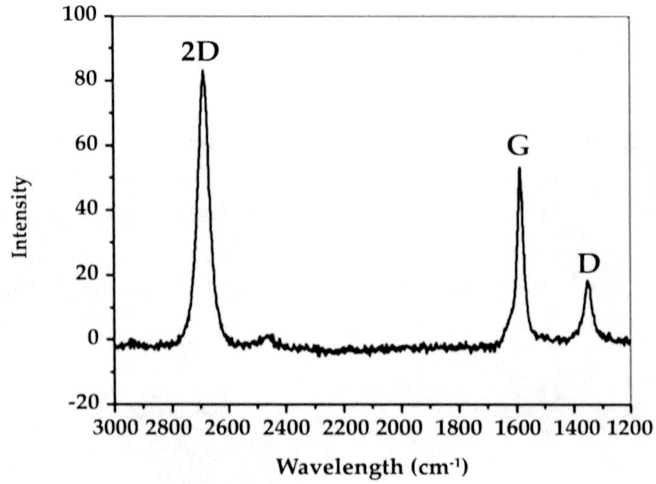

Figure 6.3. Raman spectra of CVD produced mono-layer graphene on SiO2 (300nm) /Si substrate [*Source*:https://www.researchgate.net/publication/275012245_Graphene-Quantum_Dots_Composite_for_ Photovoltaic_Devices].

The transmittance of the acquired sample can reach 95% from 400–1000 nm which also reveals the successful fabrication of single graphene layer. When QDs are attached to the surface of graphene, the transmittance decreases to almost 80–85% because of absorption by QDs. The pyridine units that are attached to the surface of the (CdSe)ZnS nanoparticles correlate with the surfaces of graphene through π-π interactions to make easy the formation of the pyridine/(CdSe)ZnS-graphene hybrids.

The absorption peak of QDs can be seen clearly for the hybrid film in Figure 6.4a. For a clear depiction of the band diagram of the device, the calculation of optical band-gap was carried out using the normalized transmittance (Guo et al., 2011; Yun et al., 2011; Ohno et al., 2013). As per the absorbance spectra, the absorption range of QDs is from 600 nm to 650 nm. Normalizing the transmittance permits us to neglect scattering and reflection losses. By making a plot of the square of the absorption coefficient, $(\alpha h\nu)^2$, against incident photon energy, hν, the optical band-gap can be obtained as shown in the Figure 6.4. For that reason, $(\alpha h\nu)^2$ has a linear dependence on hv. The value of optical energy band- gap can be obtained by extrapolating the plot's straight parts to the hv axis.

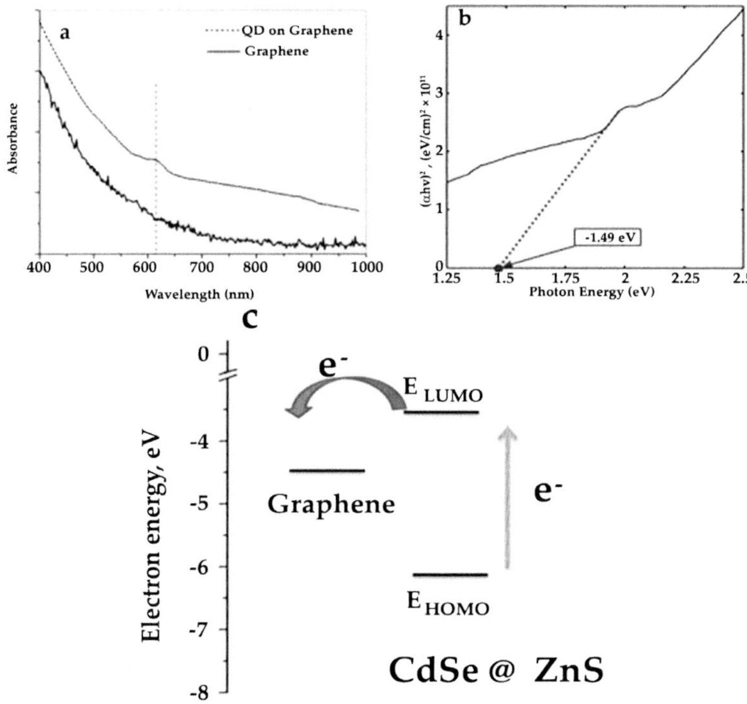

Figure 6.4. (a) UV-Visible spectra (absorption) of graphene (denoted by a solid line) and graphene-QD (denoted by dashed line). (b) Absorption coefficient square of the thin film of QD/ graphene against the incident photon energy. (c) Band illustration of graphene/ QD hybrid as calculated through cyclic voltammetry [*Source*: https://www.researchgate.net/publication/275012245_Graphene-Quantum_Dots_Composite_for_ Photovoltaic_Devices].

As reported by Kucur et al. (2007), for the calculation of LUMO and HOMO energy levels, cyclic voltammetry was used. Briefly, QDs coated with pyridine were mixed with PVDF binder, which was afterward coated on to glassy carbon electrode as a working electrode. As a counter electrode, Pt wire electrode is used in this experiment.

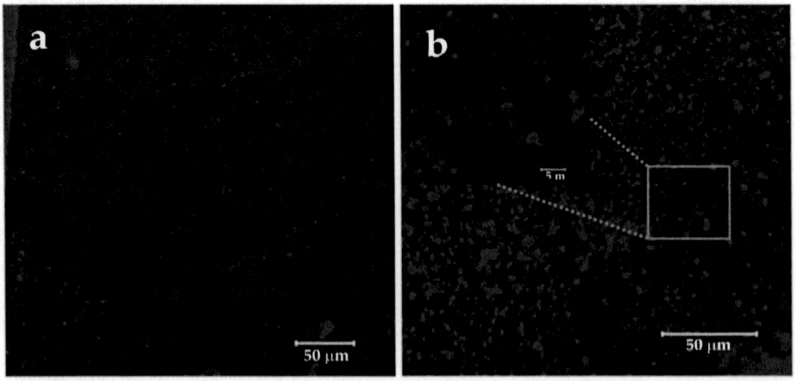

Figure 6.5. Images of con-focal fluorescence spectroscopy of graphene-QD. a) A large area of QDs on graphene with the 2x objective; b) local area with cumulative QDs with a 20x objective. The inset of b) displays the edge of glass and graphene [*Source*: https://www.researchgate.net/publication/275012245_Graphene-Quantum_Dots_Composite_for_ Photovoltaic_Devices].

The electrolyte used here for supporting was reported to be 0.1 mol/dm3 tetra-butyl-ammonium-hexafluoro-phosphate acetonitrile solution. LUMO and HOMO energy level were calculated according to the following equation:

$ELUMO = -[(E_{red} - E_{1/2(ferrocene)}) + 4.8]$ eV or $EHOMO = -[(E_{ox} - E_{1/2(ferrocene)}) + 4.8]$ eV,

where, E_{red} is reduction potential, and E_{ox} is oxidation potential.

$E_{1/2(ferrocene)} = 0.41$ eV,

$E_{red} = -0.357$ eV, $E_{ox} = 1.357$ eV, $E_{LUMO} = -4.033$ eV and $E_{HOMO} = -5.747$ eV.

Using this method, the band gap is 1.71 eV which is close to the 1.49 eV value of optical band gap. The diagram of the band was drawn following the calculation (Figure 6.4c). This provides proof that transfer of an electron from QD to graphene may be achieved with light illumination. Therefore, optimal conditions might be acquired for high-efficiency devices.

6.4.2. Distribution of QDs on Graphene

For confirmation of successful large area combination of QD and graphene, confocal fluorescence microscopy is used (Figure 6.5). It is evident that huge area of well-dispersed QDs on to graphene has been attained. Hardly any aggregated QDs are observed on graphene around the edges (Figure 6.5b). The inset noticeably demonstrates the boundary in between glass and graphene, here darker part denotes glass without QD, whereas, brighter part shows graphene with attached QDs. However; these particles present the clearer evidence of quenching effect. Itself QD has high fluorescence whereas graphene satiates the fluorescence of a single layer of QD. As graphene displays wide absorption across the visible spectral range, we can expect resonant (Förster) transfer of energy to occur from nano-crystals to single-layer graphene which induces the effect of quenching.

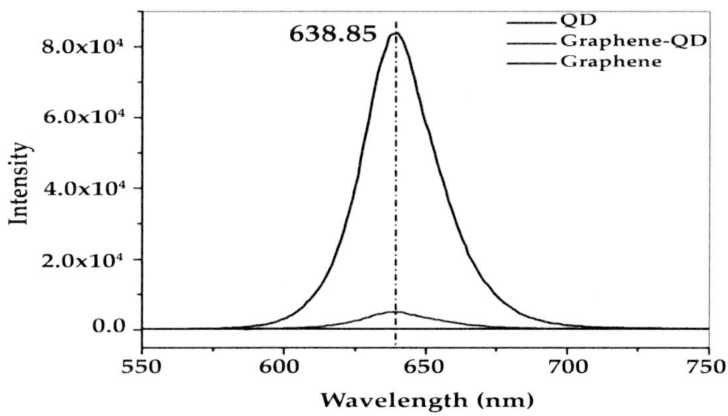

Figure 6.6. Photoluminescence spectra of Graphene/QD, QD, and Graphene samples. λex = 450 nm [*Source*:https://www.researchgate.net/publication/275012245_Graphene-Quantum_Dots_Composite_for_Photovoltaic_Devices].

For studying the quenching effect in detail, photoluminescence spectra of QD, graphene and graphene/QD samples were taken as shown in the Figure 6.6. The CdSe@CdS nanoparticle exhibits strong fluorescence in the presence of light (λ_{ex} = 450 nm), whereas, the hybrid material displays an evident fluorescence quenching effect.

After attaching on to the surface of graphene, the fluorescence intensity of QD nano-particles reduces 85%. It has been reported that the development of CdSe-SWNT hybrids favors the transfer of holes or electrons from QDs (donor) to the SWNTs (acceptor) in such a manner that the exited holes or electrons enter the SWNTs, rather than being emitted. The scenario is alike for the hybrid film in which electrons transfer takes place from (CdSe)ZnS nanoparticles to graphene, hence not emitted. The considerable decrease of fluorescence intensity implies the efficient transfer of electrons.

Besides a suitable band diagram, another significant factor of the high-efficiency device is to have a fine distribution of QDs on the graphene. QD consisting of a single layer will maximize the area of coverage as well as improve the distribution. Enhanced distribution of QDs will reduce electron-hole recombination and raise the yield of effective transfer of charge. AFM and SEM can be used for characterization of the samples. We deliberately select such an area where graphene was ragged during transfer for SEM (Figure 6.6). It is clear that QDs cover the entire graphene surface except for the triangular area.

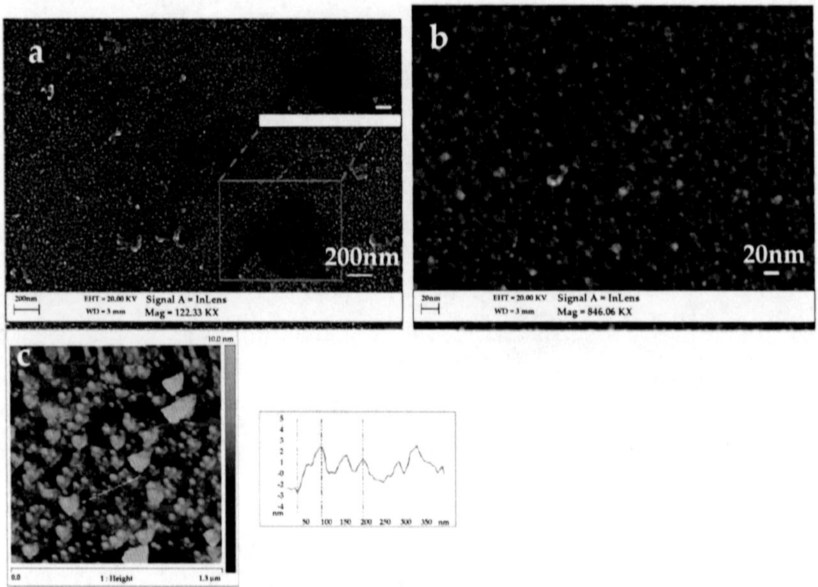

Figure 6.7. (a) SEM images, (b) SEM images of specimens, (c) AFM images of QD on the surface of graphene. [*Source*: https://www.researchgate.net/publication/275012245_Graphene-Quantum_Dots_Composite_for_ Photovoltaic_Devices].

Figure 6.7 show the inset image where no QDs can be seen in the bare glass substrate. This fact makes the success of the non-covalent bonding attachment of QDs on graphene, more clear. As can be seen in the higher magnification SEM of the sample, a single layer of QDs is well formed on the graphene surface. Still, each of the groups attaches to surface of graphene firmly without overlapping. To further ascertain the single layer structure formed on the surface of graphene device, an as-fabricated GFET can be characterized with AFM, which visibly reveals that 3–5 nm QDs were compactly spread on the surface with just a few aggregates (Figure 6.7).

6.4.3. Electronic Device

Another significant factor for designing highly efficient solar cells is a fast transfer of electrons. The as-acquired QD/graphene hybrid can be used as a field effect transistor channel for a similar study.

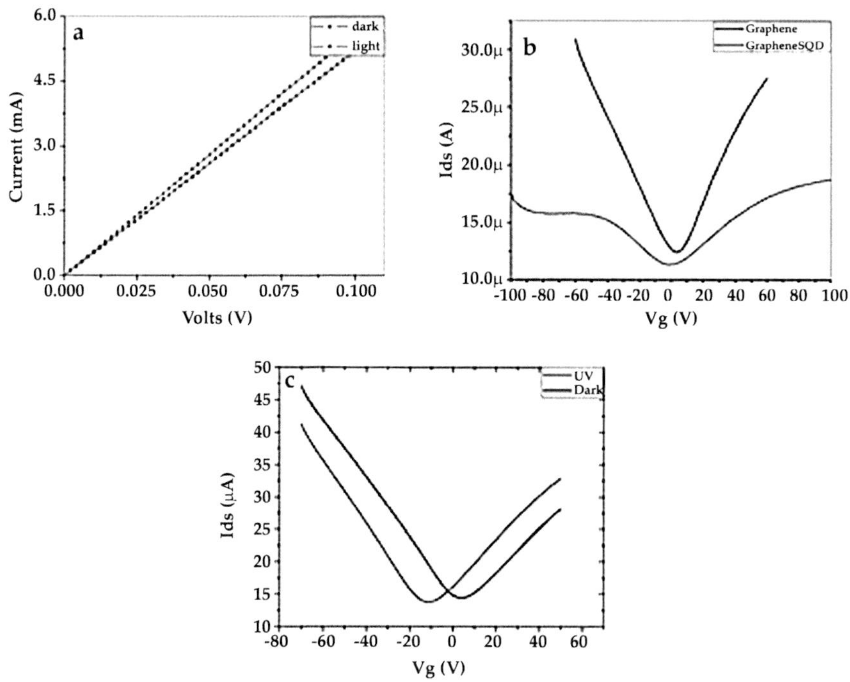

Figure 6.8. (a) Characteristic Ids-Vg curves of graphene-QD devices and graphene (c) Characteristic Ids-Vg curves of graphene-QD devices and graphene both with and without UV illumination. Bias voltage taken is 100 mV. [*Source*: https://www.researchgate.net/publication/275012245_Graphene-Quantum_Dots_Composite_for_ Photovoltaic_Devices].

Figure 6.8 displays enhancement in conductivity after UV illumination. We can monitor the resistance of the graphene/QD device by applying a 100 mV bias voltage. As graphene exhibits the property of electron-transport with UV illumination at the same time as $V_g = 0$, the decline in the resistance with the light on points out that the direction of the transfer of electrons is from the nanoparticles to graphene. During this process, the concentration of electron in graphene increases. This incident is also by the reports of CdSe electron transfer to SWNT. Fascinatingly, the mobility of the hole as well as mobility of the electron gets decreased for the hybrid material which may be due to long-range scattering. Figure S-2 in ESM represents signals of source sweeping with different applied voltages. When a positive or negative voltage is applied to GFET, it exhibits hole or electron conductivity respectively. Transfer of electron increases the concentration of carrier in electron branch whereas recombination with hole reduces the concentration of carriers in hole branch. It is evident that when we turn on UV, the conductivity of electron conductive branch is decreased whereas the conductivity of hole conductive branch increases. This behavior reveals that transfer of electrons from QDs to graphene takes place when excitons are produced. The figure shows that with illumination by UV, the Dirac Point shifted to the left (Figure 6.8) which indicates n-doping effect which further confirms the transfer of electrons from QDs to graphene. Besides transfer of electrons under illumination, the experimented Dirac point shift may also be partially accredited to a gate effect. After the process of electron transfer, the outstanding positive charges in the QD nano-particles can also have an electrostatic impact on graphene. Positive charges can drive the Fermi level (E_F) shift, which leads to the left shift of Dirac point.

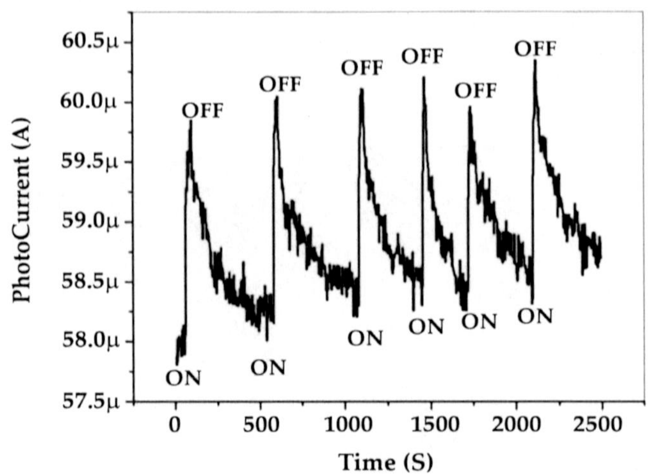

Figure 6.9. Effect of Photoswitch of graphene-QD devices under UV illumination through the application of 100 mv bias voltage. The input power of UV lamp is taken 200 mW/cm^2. [*Source*: https://www.researchgate.net/publication/275012245_Graphene-Quantum_Dots_Composite_for_ Photovoltaic_Devices].

The effect of photo-switch of the hybrid material has been tested using 100 mv bias voltage. The light source employed is a 365 nm UV light having an intensity of 200 mW/cm^2. It is evident that as the light is turned on, the resistance decreases or increases when the light is turned off (Figure 6.9). The photocurrent is generated rapidly in this process as it requires extensive time to relax to the original state. This further indicates a large lifetime of exciton owing to the effective separation of charge from the nanoparticles to the graphene. The result displays the capability of the hybrid material for the development of the future solar cell.

6.4.4. Photo-Electrochemical Cell

With the achievement of the single-layer QDs on graphene, the material should exhibit excellent photovoltaic properties with all the benefits, such as high concentration of carrier as a conductor, suitable band gap, fast transfer of electrons and high crystallinity.

For a demonstration of the prospective application of such hybrid material for solar cells, we have explained the construction of a photo-electrochemical cell (PEC). For future development of energy, PEC is one of the capable light-harvesting devices. We use QD/graphene/ITO as a working electrode, whereas, Pt is used as a counter electrode.

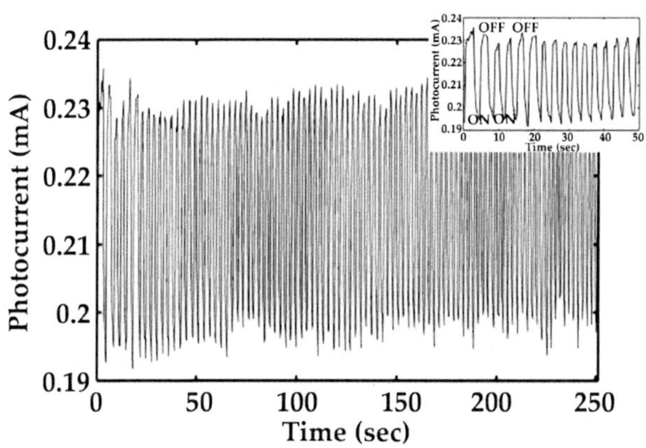

Figure 6.10. Swift short-circuit photocurrent (Isc) photo-response of QD/graphene/ITO devices under illumination of white light in a photoelectrochemical cell. The inset displays an enlarged view of a small section of the response curve of the photocurrent. Input light power is 30 mW/cm^2

[*Source*: https://www.researchgate.net/publication/275012245_Graphene-Quantum_Dots_Composite_for_ Photovoltaic_Devices].

QD/graphene/ITO device displays steady and prompt photocurrent responses. Moreover, they can be reproduced during repeated on/off cycles of illumination of light (Figure 6.10). The photo-response has a comparable result with CNT/TiO_2 hybrid material. 1-D characteristic of graphene provides it a higher surface area which presents a higher possibility for transfer of charge from 0-D materials. In comparison with 2-D (CNT)/0-D (QD) hybrid structure, hybrid structure of 1-D (graphene)/ 0-D (QD) make the possible faster transfer of electrons. Such fast response implies rapid charge transfer between graphene and QD which paves the means for the development of a solar cell anode with higher efficiency.

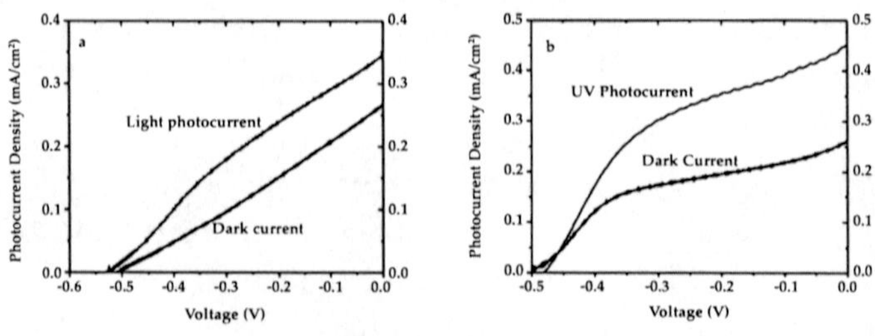

Figure 6.11. I-V characteristics of photoelectrochemical cell having linear sweep voltammograms characterized in three-electrode mode using (a) white light and (b) 365nm ultraviolet light [*Source*: https://www.researchgate.net/publication/275012245_Graphene-Quantum_Dots_Composite_for_Photovoltaic_Devices].

Figure 6.11 shows the generation of photocurrent curves. It appears that short-circuit photocurrent density (I_{sc}) increases with UV illumination or light in comparison with the dark condition. The I_{sc} can reach up to 450 μA/cm². This value is much higher as compared to previously reported CNT/QD hybrid material. The rise in UV illumination can reach 0.2 mA/cm², whereas, it still reaches 0.12 mA/cm² with quiet lower power white light illumination. According to various reports, a multilayer graphene/QD hybrid can reach higher Isc owing to larger electron transfer. In such a case, 1 layer of CVD graphene can get half of the maximum photocurrent density that is achieved with 8-layer reduced graphene/CdS; this further display enhanced the performance of the current hybrid material. IPCE with illumination by UV lamp has been calculated as 3% which is analogous to CdS/SWNT. This low approximate value of IPCE is reasonable for monolayer hybrid. Even when the lamp power is as small as 30 mW, the I_{sc} can reach 78% of the UV illumination condition. By employing multilayer of QD/graphene hybrid, we can expect higher IPCE and better transmission of light.

To summarize we can say that (CdSe)ZnS QDs can be successfully attached to surface of graphene. Efficient transfer of electrons is obtained from QDs to graphene with a single-layer dispersed QD/graphene sample. The fast to-response enables the device for photo-switch device applications.

CHAPTER 7

QUANTUM DOTS BASED HIGH-EFFICIENCY LEDS

CONTENTS

7.1. BACKGROUND

Colloidal quantum dots-based light-emitting devices (LEDs) are of vital importance for display applications which include white lighting and thin-film displays with enhanced and selectable colors (Shirasaki et al., 2013; Supran et al., 2013). 1 parameter for characterizing the performance of QD-LEDs is known as EQE (external quantum efficiency), which signifies the number of emitted photons per incident electron on a device. Recently, the red QD-LEDs having an EQE of 18% have demonstrated the potential of QD-LEDs to compete and ultimately exceed the efficiency of organic LEDs (OLEDs) (Kwak et al., 2012; Mashford et al., 2013). However, most of the QD-LEDs, especially those emitting green or blue light exhibit lower EQE values. A clear understanding of the challenges and scientific phenomena for efficiency limits is essential for the precise manufacturing of device architectures with QD chemistries for developing highly efficient QD-LEDs (Kim et al., 2008; Anikeeva et al., 2009; Qian et al., 2011).

7.2. EFFICIENCY IN QD-LEDS

Efficient electron-hole injection, decrease in the electric field across the quantum dot layer and charge carrier balance for charge carriers reaching the QD based active layer are all critical design parameters for ensuring high-performance in QD-LEDs (Anikeeva et al., 2007; 2008; Wood et al., 2009). However, these design considerations are astringently device specific and hard to accomplish in a particular device for various color emitters having different sizes and chemistries.

This chapter presents a thorough insight into the efficiency examination of a QD-LED in the last stage of the light generation process (Kuno et al., 1997; Chen et al., 2015). The probability for photonic recombination in the presence of an exciton in a QD is typically verified by using the concepts of luminescent quantum yield (QY). The EQE of a device can be implicitly assumed to be directly related to the emitters' QY in the device structure, which mainly depends on the exciton recombination rates (i.e., radiative: k_r and nonradiative: k_{nr}).

$$EQE \alpha QY = k_r / (k_{nr} + k_r). \quad (1)$$

Figure 7.1 schematically illustrates that determination of k_{nr} and k_r is carried out by the nature of QDs and their interaction with charge (Q) and the electric field (F) resulting from the current and voltage required for the operation of an LED.

Two primary contributors to non-radiative rates in a QD-LED are free-charge carriers and electronic trap states. For instance, trap-assisted recombination can take place in the presence of surface defects in the QD emitter (Klimov, 2000; Klimov et al., 2000; Kalyuzhny & Murray, 2005). On the other hand, the hole or electron in an exciton shows stability at a trap state followed by the recombination of two carriers without photonic emission. This phenomenon results in a decreased recombination rate for non-radiative trap-assisted recombination, thereby enhancing the QY (luminescence) of the QD (Galland et al., 2011; Bozyigit et al., 2013; Park et al., 2013). This result prospective has been exhibited by the presently developed "giant" shell CdS/CdSe QDs, where the CdS is coated over CdSe core in such a way that the diameter of the core is less than half of the shell thickness (Dabbousi et al., 1997; Mahler et al., 2008; Chen et al., 2013). These QDs have reportedly shown excellent sustainability at high degrees of thermal stresses along with the retention of their luminescence even after removal of the passivating ligands (Hines et al., 1996; Talapin et al., 2009; Shirasaki et al., 2013).

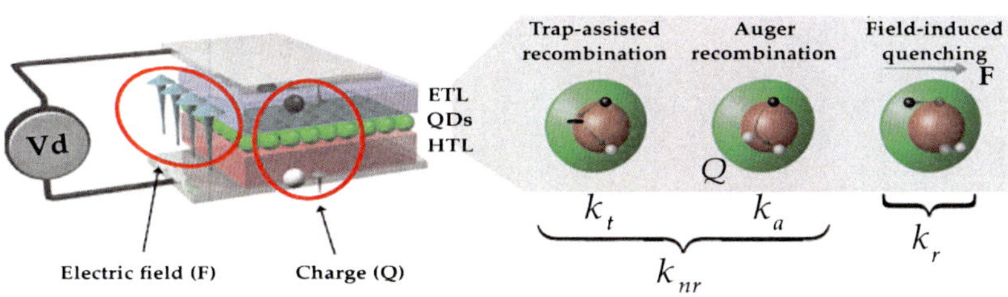

Figure 7.1. Schematic illustration of a standard QD-LED (quantum dot light-emitting device/diode) architecture containing QDs sandwiched between HTL (hole transport layer) and ETL (electron transport layer). In addition to the material defects in QDs, which may result in enhanced trap-assisted recombination rates (kt), the current and voltage required for the functioning of LEDs also result in the generation of an electric field across the QDs which can affect the radiative and non-radiative recombination rates of excitons produced in the QD-LED. [*Source*: http://sciencedocbox.com/Physics/67265475-Challenges-and-solutions-for-highefficiency-quantum-dot-based-leds.html].

The addition of shells tends to steadily enhance the QD passivation contingent to the shell material's energy levels relative to the energy levels of the core. The shell can increase $k_{nr,}$ which may result in a decrease of the QY (Chen et al., 2008; Spinicelli et al., 2009; Wang et al., 2009). On the other hand, the presence of certain charge (Q) on a QD as a consequence of an improper electron-hole injection balance can result in the increased non-radiative Auger recombination. In this case, the dissipation of energy takes place in the form of kinetic energy from charge carriers instead of photons. This phenomenon, ultimately, decreases QY and increases k_{nr} (García-Santamaría et al., 2009; 2011).

In most of the QD-LEDs, the active QD layer also faces an electric field in the orders of 1 MV/cm. Recently, experimental studies have shown that electric field in a QD device can induce spatial separation of the hole and electron wave-functions which is adequate to considerably decrease the exciton's radiative rate (Efros & Rosen, 1997; Cragg & Efros, 2009). However, k_{nr} shows no effect upon the application of an electric field in a QD device. In a case where k_r is shown smaller values (usually in the order of k_{nr} (i.e., $k_r \leq k_{nr}$), Equation 1 exhibits that a reduction in the radiative rate can result in a substantial decline in the EQE of a QD-LED (Hohng & Ha, 2004; Javaux et al., 2013).

7.3. DESIGNING QUANTUM DOTS FOR LED APPLICATIONS

Substantial developments have been made in the last 20 years in the area of quantum dot LEDs for improved quantum yields of the quantum dot emitters. These advances in chemistry can be widely classified into two trends, i.e., grading the composition of QD core and overcoating of the quantum dot core with a shell material. Figure 7.2, schematically shows that these alterations in the QD devices can result in a variation in the exciton's degree of confinement concerning a certain degree of surface passivation. This chapter addresses the impact of QD chemistry changes on the improved quantum yields of the Light emitting devices through a decline in the non-radiative processes. For instance, the shell addition results in the passivation of the QD core and offers provides physical separation between the exciton and the defect states on the QD surface (Osovsky et al., 2009; Pal et al., 2011; Bae et al., 2013).

The above mentioned phenomenon results in the decrement of trap-assisted non-radiative recombination rate (k_t), consequently improving the luminescence quantum yield of the QDs (Steckel et al., 2006). The wave functions of holes (red shading) and electrons (blue shading) which constitute the exciton are more intensely spatially confined in the core (e.g., CdSe) due to the presence of large energy offsets among the conduction and valence bands of the ZnS and CdSe (Coe-Sullivan et al., 2005; Caruge et al., 2008). This enhanced exciton confinement can be conveniently observed as the emission shift in the quantum dots to greater energies, as anticipated from a quantum mechanical phenomenon of particle-in-a-box. A typical red-emitting quantum dot of CdSe overcoated with a shell of ZnS will emit the green light. This spatial confinement stimulates strong carrier-carrier interaction that results in the retention of high non-radiative Auger recombination rates (k_a) in the core.

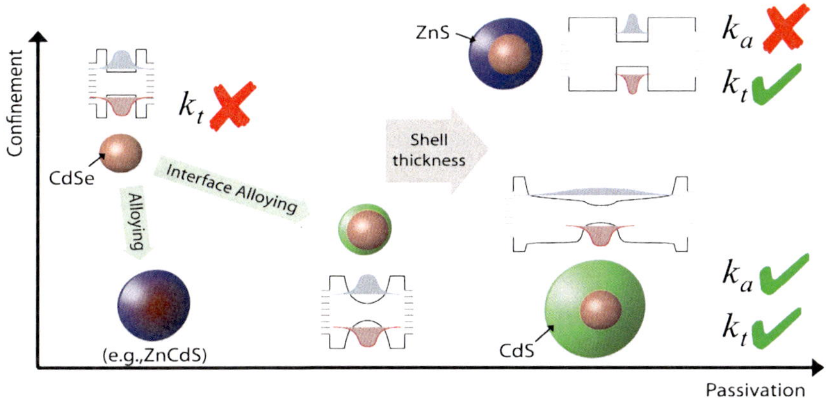

Figure 7.2. Schematic illustration of the impacts of the QD structure on the Auger (ka) or trap-assisted (kt) non-radiative recombination rates, i.e., negative impact (red-cross) or positive impact (green check). Organization of the QD structures is based on the extent of their electronic confinement and surface passivation concerning a CdSe core (present in the upper left corner of the image). ALl quantum dot structures are demonstrated along with their energy band diagrams exhibiting the wave functions of holes (red) and electrons (blue) [*Source*: http://sciencedocbox.com/Physics/67265475-Challenges-and-solutions-for-highefficiency-quantum-dot-based-leds.html].

On the other hand, when a CdSe quantum dot is overcoated with CdS, wave-function of a hole remains restricted in the CdSe core. However, the electronic wave-function outspreads into the shell (Cho et al., 2009; Voiry et al., 2013). This reduction in the electronic confinement decreases carrier-carrier interactions which ultimately results in the decreased non-radiative Auger recombination rates (k_a) (May et al., 2012). Alloyed QDs are typically considered to have a lower k value, e.g., ZnCdSe or ZnCdS (19). This disparity has been theoretically explained by smoothing the confinement potential's shape (see the lower left corner of Figure 7.2) which is assumed to occur in these alloyed quantum dots (Bozyigit et al., 2013; Nam et al.,

2015). Alloying is also assumed to be present at the interface (shell-core interface) of the "giant" shell quantum dots, which considerably explains the reasons of lower K_a values apart from the reduced electron confinement in these alloyed QDs.

The above mentioned considerations suggest that "giant" shell quantum dots are an ideal choice for developing an emitter of a QD-LED because of their optimal passivation and reduced non-radiative Auger recombination rates. However, QD-LEDs manufactured from "giant" shell CdS/CdSe QD materials have not shown extraordinary efficiencies (Muller et al., 2005; Kraus et al., 2007). On the other hand, QD-LEDs having multilayered alloyed architectures (e.g., ZnCdSe/ZnCdS, CdZnS/CdZnSe, etc.) display the highest EQEs. This trend can be explained by taking into account the effect of the electric field on the quantum yield of various QD types (Empedocles et al., 1996; Park et al., 2007).

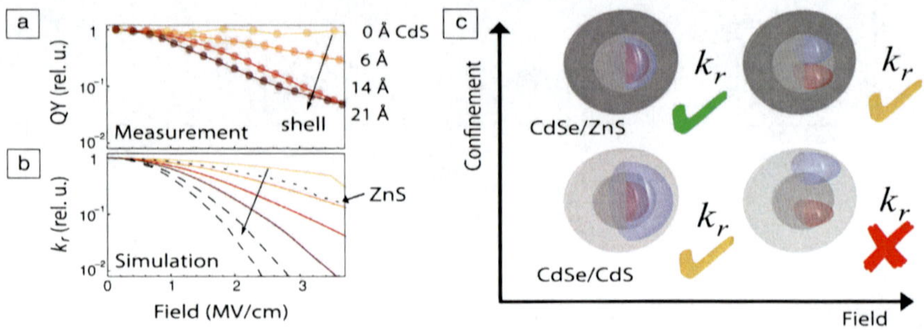

Figure 7.3. (a) Experimental measurements for the photo-luminescent QY (quantum yield) and (b)calculations for tight binding of the elements in an optical matrix concerning the electric field (F) experienced by the QDs. (c) Schematic demonstrations for wave-functions of holes and electrons in CdS/CdSe and CdS/ZnSe QDs with and without the stimulus of an electric field which shows that relaxed confinement is undesirable in the existence of an electric field, as it causes a decline in the radiative rates because of the exciton polarization. [*Source*: http://sciencedocbox.com/Physics/67265475-Challenges-and-solutions-for-highefficiency-quantum-dot-based-leds.html].

Figure 7.3a show that the increase in the electric field across CdS/CdSe QDs results in the decreased luminescence quantum yield (QY). The degree of the QY decrease that can reach up to 10 folds of the electric fields applied to a QD-LED is usually regulated by the thickness of the shell (e.g., CdS). ZnS shell exhibits significantly small quenching; nevertheless, it is still present in the device because of its importance in the efficiency of the

QD-LEDs. Recently, an experimental study has revealed that the efficiency drops in QD-LEDs (as observed in current density vs. EQE plots) is caused by the field-induced luminescence quenching (Empedocles & Bawendi, 1997; Bozyigit et al., 2013).

The luminescence quenching induced by the electric field has been witnessed for QDs and several other nanocrystalline architectures (Rothenberg et al., 2005). Non-radiative Auger recombination suggests that the origin of luminescence quenching (driven by the electric field) is closely associated with QD band architecture. Time-dependent measurements of photoluminescence in the QDs (being exposed to variable electric fields) unveil that lifetimes of excitons are invariant concerning the field. This indicates that the non-radiative rates remain unaltered with the application of the electric field. Calculations carried out for optical matrix element (Figure 7.3b), which is directly related to the radiative rates (Kr) and typically reflects the spatial overlaps in the wave functions of electrons and holes, confirm that the decline in luminescence quantum yield be associated with the decrement in the radiative rates (Jarosz et al., 2004). Figure 7.3c schematically illustrates that the selection of a QD chemistry with lower electronic confinement (e.g., a CdSe core QD with CdS shell in place of a ZnS shell) permits for the spatial parting of the wave functions for holes and electrons (i.e., red and blue shading) when QDs are exposed to the electric field. This particular effect can be regarded as exciton polarization which results in the reduced radiative recombination rates (kr) (Figure 7.4)

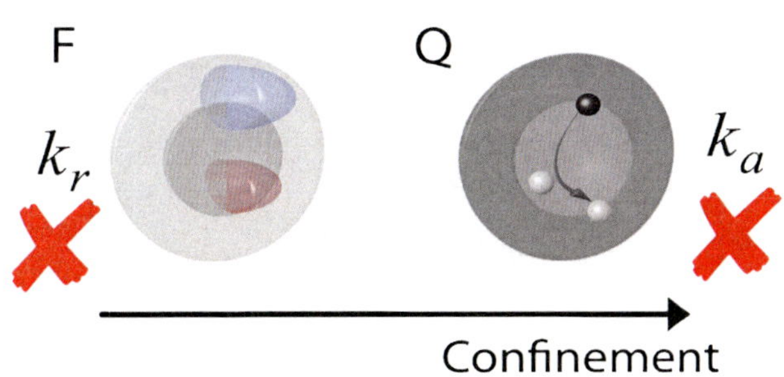

Figure 7.4. Schematic illustration of the challenge encountered in the selection of a quantum dot emitter for LED (light-emitting diode) applications. QDs with lower electronic confinement possess low non-radiative Auger recombination, but also show decreased radiative recombination in the existence of an electric field. On the other hand, a quantum dot with high electronic confinement exhibits high non-radiative Auger recombination and lower luminescence quenching induced by an electric field.

[*Source*: http://sciencedocbox.com/Physics/67265475-Challenges-and-solutions-for-highefficiency-quantum-dot-based-leds.html].

7.4. SOLUTIONS FOR ACHIEVING HIGH-EFFICIENCY QUANTUM DOT-LEDS

Design of a QD emitter plays a central role in defining the performance of a QD-LED. Literature has shown that a multilayered emitter architecture with following characteristics exhibits superior performances:

- A core having smoothed confinement potential for achieving minimized non-radiative Auger recombination is required for high-performance QD LEDs.
- A shell is required which confines the electron-hole wave functions in a core in such a way that it passivates the QD and minimizes the exciton polarization in the presence of an electric field.

However, the synthesis of such QD structures can be challenging for a range of desired wavelengths. Additionally, a particular band structure for each light-emitting quantum dot in the QD-LEDs would require tailoring to the local charge distributions and electric fields present in the quantum dot during the operation of a QD based LED device (Tani et al., 2008).

7.4.1. Luminescence of Quantum Dots from Localized State

One approach to enhance the QD emitters' performance is to minimize the luminescence quenching (field-induced) in quantum dots already displaying high photo-luminescent quantum yields (QYs) and low non-radiative Auger recombination rates (Kobayashi et al., 2007; Wood et al., 2009; Wood & Bulović, 2010).

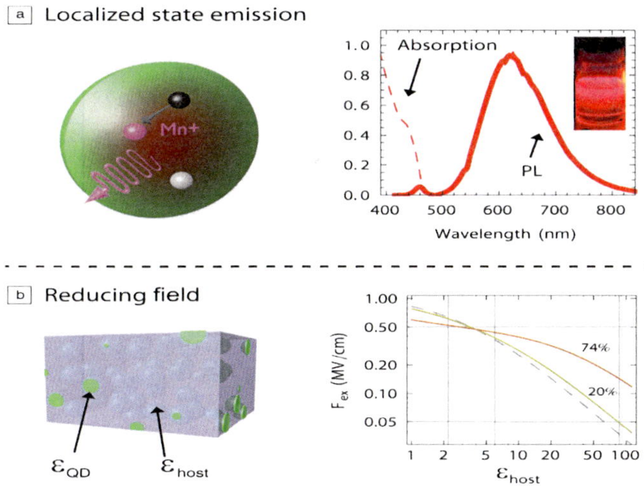

Figure 7.5. Two possible approaches to circumvent luminescence quenching caused due to the electric field. (a) A dopant has the potential to provide a luminescent state which can be less sensitive to higher electric fields. Mn+ is 1 common type of dopant in QDs which emits the red light (solid line) even if the parent QD has a high bandgap, as shown in the absorption onset (dashed line) at 450 nm. (b) Embedding quantum dots in a host (parent) material with high dielectric constant (ε) can significantly diminish the effective electric field experienced by the excitons (Fex). For instance, an emissive layer containing a TiO2 matrix with a 20% quantum dot volume fraction decreases the effective electric field experienced by the quantum dots by a factor of 10 (or slightly more), as compared to a pure QD film. [*Source*: http://sciencedocbox.com/Physics/67265475-Challenges-and-solutions-for-highefficiency-quantum-dot-based-leds.html].

One aspect of this approach may involve the fast transport of an exciton generated in a thick-shelled or alloyed QD to a localized state in a way that exciton polarization is reduced in the existence of an electric field (Chung et al., 2001; Porter et al., 2008). Minimization of the luminescence quenching induced by the electric field can also be achieved via the incorporation of a luminescent impurity (e.g., Mn) $^{+}$ into the alloyed quantum dot (See Figure 7.5a). Moreover, the selection of a quantum dot can also play a role in achieving the desired results, i.e., selection of a QD (e.g., CuInSe or CuInS) which involves the occurrence of luminescence through a donor-acceptor route (Thakar et al., 2007; Aldakov et al., 2013; Kolny-Olesiak & Weller, 2013).

7.4.2. Placement and Engineering of QD Active Layer

Another method to compensate the limitations in EQE due to the presence of luminescence quenching induced by electric field involves the minimization of the high-electric-field for the exposed QDs (Pourret et al., 2009; Lambert et al., 2010; Leschkies et al., 2010). As a primary step, this suggests that the placement of QDs should be made away from the material interface, where a high electric field is typically present (~ 1 MV/cm). It has been experimentally proved that this approach can substantially improve the efficiencies of QD-LEDs (Kinder et al., 2011; Liu et al., 2011; Moroz et al., 2013).

CHAPTER 8

PROSPECTS AND ACCOMPLISHMENTS OF QUANTUM DOTS IN PHOTOVOLTAICS

CONTENTS

8.1. INTRODUCTION

The significant breakthrough for the development of defect-free quantum dots (QDs) was marked by the exploration of the Stranski-Krastanow growth of stressed semiconductor heterostructures. For the very first time, single QDs are enabling us to develop the electrically functioned emitters of single-polarized or entangled photons: a vital constituent for quantum communication systems. QDs that are injected in quantum wells and stacked in planes above each other have directed us to develop such semiconductor lasers which can operate at wavelengths that were formerly impossible, or at least hard to reach, for a given family of the semiconductor. These lasers exhibit large radiation resistance, excellent temperature stability, and excellent spatial and temporal stability. In this chapter, we will discuss current progress in the field of quantum dot devices in photovoltaic (Stier et al., 1999).

In semiconductors, the electronic density of states (DOS) of charge carriers is considerably modified if the wave function is confined spatially. Levels of energy are quantized in all spatial directions in the final limit of three-dimensional charge carrier confinement, and there are discrete energy levels. For specific spatial dimensions of around ~10 nm, the energy separation of the levels lies within the range of 10–100 meV, subjective to the effective carrier mass. In such a single quantum dot, the discrete energy spectrum is similar to that of an atom embedded in a dielectric matrix supplied by a confining semiconductor with a larger band gap (Bimberg et al., 1999).

8.2. HISTORY AND BACKGROUND

During the 1970s, Dingle and Henry (1976) found that quantization of energy in quantum wells can be exploited for semiconductor lasers for lowering the lasing threshold as well as for providing a tunable wavelength (Tsang et al., 1979). Size quantization results in an increase of the DOS nearby band edge in QWs. When we further reduce the dimensionality of the active layer, inserted charge carriers concentrate in a progressively narrow energy range close to the band edge. As a result, the maximum material gain upsurges and the dependence of laser parameters upon temperature is reduced. In the 1980s, these effects were already predicted theoretically by Arakawa & Sakaki (1982; 1983) for lattice-matched heterostructures such as GaAs/AlGaAs (Asada, 1915; Hirschman et al., 1997). By taking into consideration, the dots having a cubic shape and containing merely one confined hole and electron state, gain increase and a corresponding decrease in the threshold current was predicted, together with an increase of temperature stability. The concluding point is a striking illustration of the effect of a discrete DOS. Although in bulk materials, when the temperature gets higher, charge carriers are continuously distributed to higher states of energy, whereas, no such redistribution can take place in zero-dimensional structures if the excited states are present sufficiently above the ground state. At that time, all carriers, independent of temperature, contribute to the inversion at lasing energy (Figure 8.1).

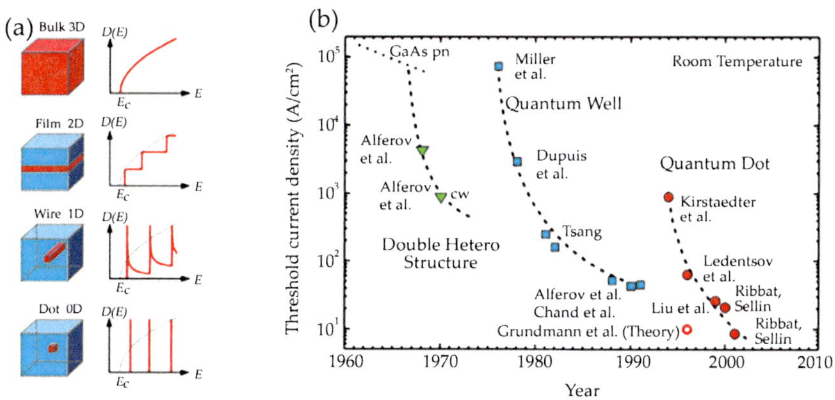

Figure 8.1. Impact of Reduced Dimensionality (a) Charge carriers is restrained in a semiconductor (denoted by red) having a smaller bandgap than the cladding semiconductor (denoted by blue). The confinement alters the density of states D(E). Ec is the conduction-band edge of a 3D semiconductor. (b) Threshold current-density development of heterostructure lasers. A noticeable decrease in threshold can be accomplished by introducing structures having reduced dimensionality: quantum wells and quantum dots. [*Source*: https://www.sciencedirect.com/science/article/pii/S1369702111701833].

These lasers were not built on lattice-matched hetero-structures, but they are based on strongly difficult ones. This breakthrough became successful when the self-organized progress of dense arrays of coherent quantum dots, based on strained-layer hetero-epitaxy with a large band-gap barrier material, was developed. The first quantum-dot laser that was developed using this approach verified the anticipated ultra-low threshold current density (120 A/cm^2) as well as its low-temperature dependence (T_0 = 350 K) at 77 K (Kirstaedter et al., 1994; Kamath et al., 1996). In the coming years, rapid improvements were made in the performance of quantum-dot photovoltaic (lasers), as demonstrated by the current threshold data (Figure 8.1b) (Ledentsov et al., 1994; 2000). Thereby introducing quantum-dot lasers to the market at the beginning of the new millennium (Grundmann, 1999; Grundmann et al., 1999).

The initial guesses of enhanced performance of quantum dot lasers were mostly based on suppositions, which later were observed not to apply. Usually, real quantum dots have various excited states which might be occupied thermally. Self-organized growth incites some variation in the shape and the size of the dots, which leads to a spread of localization energy inside a dot ensemble. The ensemble luminescence's line shape is composed of a superposition of various Lorentzian single-dot line-shapes. It is then in-homogeneously made wider (normally $\geq$ 50 meV). Therefore, the function of carrier distribution of a real dot ensemble gradually alters at high temperatures (normally $\geq$ 300 K) from a non-thermal distribution to a Fermi function (Kim et al., 2009; Bimberg & Pohl, 2011).

Device applications which are based on quantum dots may either involve a huge number of dots in an ensemble, or individual dots or just a few dots. We will first focus on applications that require a large density of dots.

8.3. DEVICES EMPLOYING ENSEMBLES OF QUANTUM DOTS

Self-organized growth can easily succumb, in a single layer, a high density of mid-10^{10} cm^{-2} quantum dots (Oshima et al., 2008). Most well-known applications of dense ensembles of quantum dots in photovoltaic are near-infrared lasers, although QDs are also quite attractive for improving other devices, for example, solar cells: previously, QDs were used for the realization of the concept of intermediate-band solar cells, where the moderately occupied intermediate band of the confined QD levels leads to additional sub-bandgap absorption in a solo-junction cell (Luque & Martí, 1997; Marti et al., 2006; Popescu et al., 2008). The proof of concept has already been demonstrated, and development in this exciting field is encouraging (Hubbard et al., 2008).

In laser applications, a large material obtains results from a high dot density, since the optical gain is directly proportional to the number of involved inverted states (Kirstaedter et al., 1996; Schmidt et al., 1996). Currently, 2 types of lasers lead the market: surface-emitting lasers and edge-emitting lasers. At first, we will focus on lasers with edge-emitting geometry.

Employing quantum dots for the active region in optical amplifiers and edge-emitting lasers present some certain benefits (Snowton et al., 2002; Ribbat et al., 2001; 2003). In case of GaAs-based devices, tuning capacity of the emission wavelength by controlling their composition and size permits operation in the significant datacom O-band (1.26 – 1.36 μm). The confinement of carrier in the dots suppresses carrier diffusion in an efficient manner (Berg et al., 2001). This results in an excellent spatial and temporal stability of the near region of the transverse ground mode, repressing filamentation as well as improving degradation resistance under high-energy particle radiation (Piya et al., 2000; Gehrig et al., 2004; Thompson et al., 2005). Experiments prove that QD lasers are robust against optical feedback into the cavity, decreasing high-frequency noise (Huyet et al., 2004; Schneider et al., 2005). Additionally, scientists have observed ultrafast population dynamics in QD-based gain media, which makes such materials interesting for applications in optical amplifiers and saturable absorbers (Borri et al., 2000; 2006).

8.3.1. Mode-Locked Lasers

Monolithic semiconductor mode-locked lasers (MLL) are in high demand for generating high-frequency optical pulse trains as sources of the pulse for optical clocks in electronic circuits, datacom, metrology, and optical sampling of high-speed signals. The principle of operation illustrated in the following figure shows how the formation of laser cavity takes place by the external facets of a ridge-waveguide structure; this structure is divided into two sections, namely, a gain section and other being a reverse-biased absorber section. Pulse trains are formed via intensity-dependent absorption in the modulated or passive (®hybrid MLL) absorber. One crucial necessity for attaining a small pulse width in mode-locked lasers is the fast recovery of the absorber (Avrutin et al., 2000; Williams et al., 2004). High-frequency pulses need small cavities for the duration of short round trip (for instance, for a passive QD device having a length of 450 + 50 μm, 80 GHz was demonstrated), and therefore, high gain and fast recovery (Kuntz et al., 2007). For most application, a small timing jitter inside the optical pulse train is of significance. The decreased feedback sensitivity and ultrafast gain recovery of gain media based on QD promises a lower jitter than anticipated for QW-based devices. Outstanding data of a hybrid mode-locked device is shown in Figure 8.2b. At 40 GHz operation, a root-mean-square timing jitter of value as low as 190 fs is accomplished (Recombination, 2001). The observation illustrates that for amplified reverse bias at the absorber section, the jitter is reduced. The values observed are quite lower than the ITU recommended values for 40 GHz sources that indicate a root-mean-square timing jitter lower than 2.5 ps for the integration range of 20 MHz – 320 MHz (Schmeckebier et al., 2010). Figure 8.2c (left) illustrates transmission of data through an externally synchronized modulation of the laser pulse-train at 40 Gb/s (Meuer et al., 2009). Further, the 40 GHz pulse train has also been shown appropriate to generate optical data streams at 160 Gb/s. Application of chirp compression results in 700 fs pulses with a full width at half maximum (FWHM). This value is close to the theoretical limit. Figure 8.2c (right) exhibits a 160 GHz pulse comb produced via two-stage optical time-division multiplexing (abbreviated as OTDM) of the hybrid mode-locked QD laser.

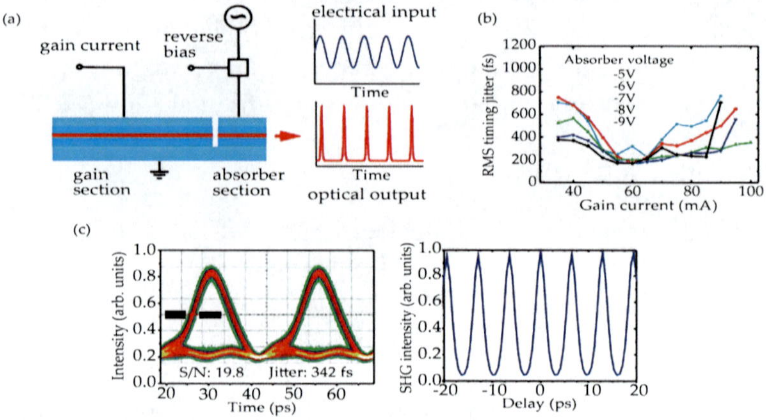

Figure 8.2. (a) Representation of a mode-locked laser having hybrid mode locking. The structure of ridge-waveguide laser is divided into gain section and absorber sections; these sections are separated by a small air gap. Asynchronous optical pulse train is generated by the modulated absorption (b) RMS timing jitter of a hybrid mode-locked QD laser being operated at 40 GHz, and derivative of the single-sideband noise integrated from 1 kHz to 1 GHz. (c) Left: External modulation of the optical pulse train of 40 GHz generates an optical data stream of 40 Gb/s. Right: Measurement of the autocorrelation of the 40 GHz pulse train, up-converted by two-stage optical-time-division multiplexing, with a pulse comb at 160 GHz. [*Source*: https://www.sciencedirect.com/science/article/pii/S1369702111701833].

8.3.2. Semiconductor Optical Amplifiers

It is expected that future optical networks will generate a large requirement for low-cost switches, linear amplifiers, optical signal regenerators, wavelength converters, and stimulating R+D of semiconductor optical amplifiers (SOAs). Devices like these consist of a ridge-waveguide structure composed of non-reflecting cavity facets (anti-reflection tilted or coated) to avoid lasing as can be seen in the scheme Figure 8.3a. For any of the above applications, the broad bandwidth and ultrafast gain recovery of QD gain media result in more benefits as compared to bulk media and QW.

Figure 8.3a shows the conversion of wavelength needed for all-optical multi-wavelength networks. For accomplishing this conversion, we use nonlinear cross-gain modulation (XGM). The SOA is intensely saturated with a large amount of current. A modulated optical signal at λ_{red} results in the modulation of the SOA gain. As a result, at λ_{blue}, gain modulation affects an additional weak continuous wave optical probe signal and hence leaves the SOA with a modulation that is inverted concerning that of λ_{red}.

Figure 8.3b compares the cross-gain modulation bandwidth of a QD SOA with a conventional QW SOA (Joergensen et al., 1997). The QW SOA' 3 dB XGM bandwidth is nearly independent of the injection current and is around 15 GHz (Kim et al., 2009). In comparison, for rising current, QD SOAs exhibit nearly frequency-independent gain well afar from 40 GHz which is quite larger than the best-reported values.

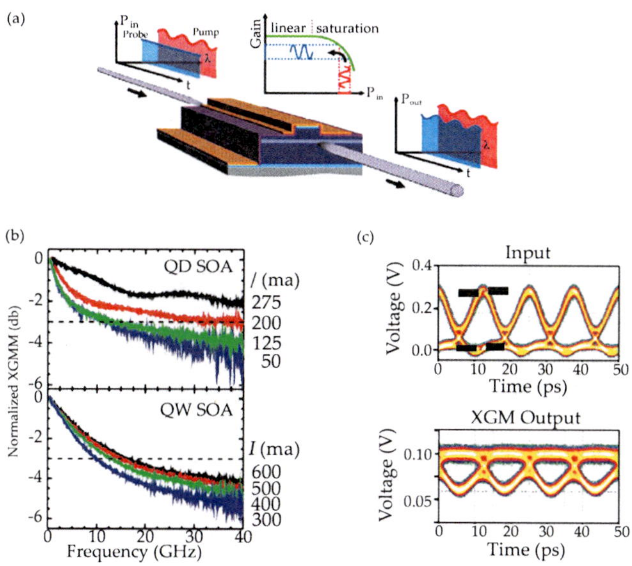

Figure 8.3. Cross-gain modulation of a semiconductor optical amplifier (SOA) (a) Principle of Operation: a strong and modulated pump signal (Pin, denoted

by red) impels the SOA into the saturation regime. As a result, the SOA gain is also modulated. The pattern of modulation can be shifted to a weak CW probe pulse at some other wavelength (Pin, denoted by blue), resulting into an inverted pattern of modulation on the probe signal in the output (Pout, indicated by blue). (b) The efficiency of the cross-gain modulation of a QD SOA (shown at top) and a conventional QW SOA (shown at bottom). The 3dB bandwidth of the QD SOA denoted by the horizontal line can be tuned at high injection current to beyond 40 GHz (c) Wavelength up-conversion by cross-gain modulation of a QD SOA and a pseudo-random RZ-OOK-modulation. Top figure: input eye illustration of an 80 Gb/s signal at a wavelength of 1292 nm, bottom figure: up-converted output signal at a wavelength of 1300 nm. [*Source*: https://www.sciencedirect.com/science/article/pii/S1369702111701833].

The difference in dependence on injection current is due to the transition among two different gain saturation methods in the QD SOA(Vallaitis et al., 2000). At frequencies lower than 10 GHz, the carrier reservoir (principally the wetting-layer QW) follows the modulation created by total carrier depletion. This method governs the QW SOA as well. Above 10 GHz frequency, spectral hole burning as well as its ultrafast recovery turns out to be the primary saturation process in the QD SOA and allows a large XGM bandwidth (Schmeckebier et al., 2010). Hence, QD SOAs take advantage from the dynamics of a QD ensemble.

Figure 8.3c demonstrates up-conversion of an 80 Gb/s data stream by using XGM of a QD SOA at 1292 nm to 1300 nm (Contestabile et al., 2010; Meuer et al., 2011). Converted output signal's eye diagram is visibly open and exhibits low patterning effects. Lately, error-free XGM has been shown between 1310 and 1320 nm at 80 Gb/s with a bit-error-rate lower than 10^{-9} (Mutig et al., 2008; 2009). This data shows the capability of XGM in GaAs-based QD SOAs for transmission of 100 Gb/s at a single-wavelength channel [31].

8.3.3. Vertical-Cavity Surface-Emitting Lasers

Semiconductor lasers having a vertical cavity presents some beneficial properties. Such lasers produce a high-quality circular beam which gets readily coupled to a fiber, functions at low threshold currents allowing high-speed modulation, and feature small temperature dependence approximately up to 85°C, and more.

Furthermore, vertical-cavity-surface-emitting lasers (VCSELs) are created in a planar technology, which permits efficient on-chip testing at an initial stage during the production process as well as the vertical integration of components such as modulators. Figure 8.4a shows the configuration of a VCSEL. The active medium (QWs or QDs) is placed in a vertical cavity which in general is formed by distributed-Bragg-reflector mirrors. At each reflection, low losses are crucial, as the cavity is very short, and the gain is low per cycle. Owing to the large output aperture, the beam divergence is extremely low. Figure 8.4b represents a completely processed GaAs-based QD VCSEL. The top-view micrograph displays the circular output aperture and the *n*-and-*p* contacts.

GaAs/AlAs-based VCSELs take advantage of higher thermal conductivity and a larger refractive-index step as compared to InP-based devices. Today, quantum-well VCSELs functioning at 0.85 μm emission (the typical wavelength in the data communication industry), are an established and a popular product in the market for optical computer mice. Significant efforts are being directed to replace copper wires in short-range (< 300 m), high-speed data links using multi-mode optical fibers and such QW VCSELs. Presently, datacom QW VCSELs for commercial purposes are designed for 10 GB/s data rates.

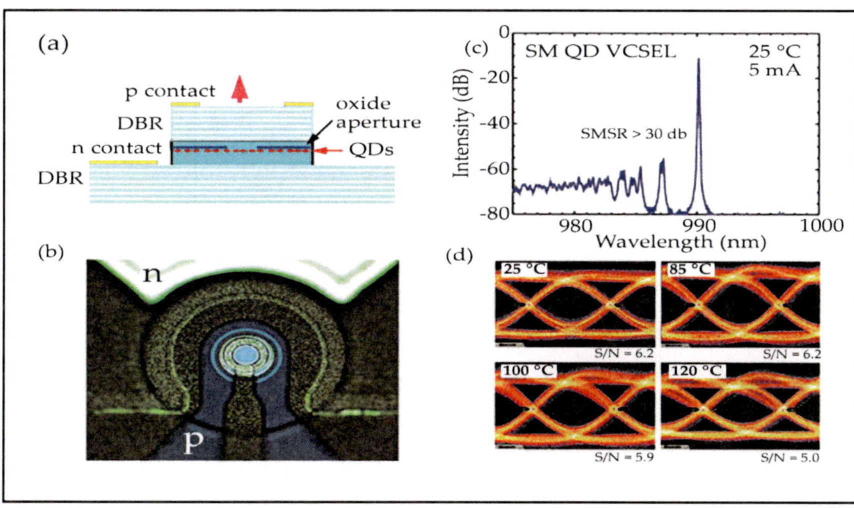

Figure 8.4. A vertical-cavity surface-emitting semiconductor laser (VCSEL) is shown (a) Cross section view of Schematic. The optically active medium containing the quantum dots (QDs) is clad with the help of 2 distributed-Bragg reflectors (DBR) creating an optical cavity. The radiation is emitted vertically (indicated by red arrow). (b) Top view of a particular device displaying the aperture and the contacts. (c) Emission spectrum displayed on a logarithmic scale.

(d) Eye diagrams are presenting 20 GB/s device's operation up to 120°C for large signal modulation. The open eyes show good signal-to-noise ratio at high operation temperature as well.[*Source*: https://www.sciencedirect.com/science/article/pii/S1369702111701833].

Current developments have achieved bandwidths greater than 20 GHz as well as large signal error-free data rates (BER less than 10^{-12}) at 40 GB/s for this wavelength (Westbergh et al., 2010; Germann et al., 2010; 2012). Recently, an error-free data rate around 44 Gb/s has been achieved for devices with strained QWs functioning at 985 nm (Saito et al., 1996; Hopfer et al., 2006; 2007). Numerous applications of quantum-dot gain media have also opened doors for 1.3 μm datacom window for GaAs-based VCSELs. The first ever developed QD VCSELs presented some time ago; it was fabricated by employing MOVPE or MBE (Lott et al., 1997; Schur et al., 1997; Lenz et al., 2010). The latest advancements have also been achieved by employing sub-monolayer quantum dots in place of the Stranski-Krastanow quantum dots introduced earlier (Hudgings et al., 1998; Ledentsov et al., 2007; 2017). Error-free data transmission was achieved at 20 Gb/s for directly modulated devices functioning at 980 nm up to 85°C, and evidently open eyes were demonstrated at this data rate up to 120°C (Shchukin et al., 2008; Hill et al., 2009).

Presently, two novel concepts of vertical emitters are emerging for developing future optical interconnects micro/ nano-lasers with a metal-coated cavity and secondly, VCSELs with an implemented electro-optically modulator (EOM-VCSELs). Proofs-of-concept have recently been provided for both approaches using quantum-well gain media (Kaiser & Kezdy, 1984; Miyazaki & Kurokawa, 2006; Lu et al., 2010).

BIBLIOGRAPHY

1. Aerts, M., Bielewicz, T., Klinke, C., Grozema, F. C., Houtepen, A. J., Schins, J. M., & Siebbeles, L. D. (2014). Highly efficient carrier multiplication in PbS nanosheets. *Nature Communications*, *5*, 3789.
2. Aerts, M., Spoor, F. C., Grozema, F. C., Houtepen, A. J., Schins, J. M., & Siebbeles, L. D. (2013). Cooling and Auger recombination of charges in PbSe nanorods: crossover from cubic to bimolecular decay. *Nano Letters*, *13*(9), 4380–4386.
3. Al-Ibrahim, M., Roth, H. K., Schroedner, M., Konkin, A., Zhokhavets, U., Gobsch, G., & Sensfuss, S. (2005). The influence of the optoelectronic properties of poly (3-alkylthiophenes) on the device parameters in flexible polymer solar cells. *Organic Electronics*, *6*(2), 65–77.
4. Al-Kuhaili, M. F. (2004). Optical properties of hafnium oxide thin films and their application in energy-efficient windows. *Optical Materials*, *27*(3), 383–387.
5. Alaferdov, A. V., Balashov, S. M., Canesqui, M. A., Parada, S., Danilov, Y. A., & Moshkalev, S. A. (2014). Formation of thin, flexible, conducting films composed of multilayer graphene. *Bulletin of the Russian Academy of Sciences: Physics*, *78*(12), 1357–1361.
6. Aldakov, D., Lefrançois, A., & Reiss, P. (2013). Ternary and quaternary metal chalcogenide nanocrystals: synthesis, properties and applications. *Journal of Materials Chemistry C*, *1*(24), 3756–3776.
7. Aliberti, P., Shrestha, S. K., Teuscher, R., Zhang, B., Green, M. A., & Conibeer, G. J. (2010). Study of silicon quantum dots in a SiO2 matrix

for energy selective contacts applications. *Solar Energy Materials and Solar Cells*, *94*(11), 1936–1941.

8. Alivisatos, A. P. (1996). Perspectives on the physical chemistry of semiconductor nanocrystals. *The Journal of Physical Chemistry*, *100*(31), 13226–13239.
9. Alivisatos, A. P. (1996). Semiconductor clusters, nanocrystals, and quantum dots. *Science*, *271*(5251), 933–937.
10. Allan, G., & Delerue, C. (2006). Role of impact ionization in multiple exciton generation in PbSe nanocrystals. *Physical Review B*, *73*(20), 205423.
11. Allan, G., & Delerue, C. (2009). Fast relaxation of hot carriers by impact ionization in semiconductor nanocrystals: role of defects. *Physical Review B*, *79*(19), 195324.
12. Amassian, A., Asbury, J. B., & Sargent, E. H. (2011). Colloidal-quantum-dot photovoltaics using atomic-ligand passivation.
13. Amato, G., & Brunetto, N. (1996). Porous silicon via freeze drying. *Materials Letters*, *26*(6), 295–298.
14. Amato, G., Delerue, C., & VonBardeleben, H. J. (Eds.). (1998). *Structural and Optical Properties Of Porous Silicon Nanostructures* (Vol. 5). CRC Press.
15. Anderson, N. A., & Lian, T. (2005). Ultrafast electron transfer at the molecule-semiconductor nanoparticle interface. *Annu. Rev. Phys. Chem.*, *56*, 491–519.
16. Anikeeva, P. O., Halpert, J. E., Bawendi, M. G., & Bulovic, V. (2009). Quantum dot light-emitting devices with electroluminescence tunable over the entire visible spectrum. *Nano Letters*, *9*(7), 2532–2536.
17. Anikeeva, P. O., Halpert, J. E., Bawendi, M. G., & Bulović, V. (2007). Electroluminescence from a mixed red− green− blue colloidal quantum dot monolayer. *Nano Letters*, *7*(8), 2196–2200.
18. Anikeeva, P. O., Madigan, C. F., Halpert, J. E., Bawendi, M. G., & Bulović, V. (2008). Electronic and excitonic processes in light-emitting devices based on organic materials and colloidal quantum dots. *Physical Review B*, *78*(8), 085434.
19. Anilkumar, P., Wang, X., Cao, L., Sahu, S., Liu, J. H., Wang, P., & Sun, Y. P. (2011). Toward quantitatively fluorescent carbon-based "quantum" dots. *Nanoscale*, *3*(5), 2023–2027.
20. Anopchenko, A., Marconi, A., Moser, E., Prezioso, S., Wang, M., Pavesi,

L., & Bellutti, P. (2009). Low-voltage onset of electroluminescence in nanocrystalline-Si/SiO 2 multilayers. *Journal of Applied Physics*, *106*(3), 033104.

21. Antončík, E. (1957). On the theory of the spectral dependence of the quantum efficiency of homopolar crystals. *Czechoslovak Journal of Physics*, *7*(6), 674–689.
22. Antonova, I. V., Gulyaev, M., Savir, E., Jedrzejewski, J., & Balberg, I. (2008). Charge storage, photoluminescence, and cluster statistics in ensembles of Si quantum dots. *Physical Review B*, *77*(12), 125318.
23. Arakawa, Y., & Sakaki, H. (1982). Multidimensional quantum well laser and temperature dependence of its threshold current. *Applied Physics Letters*, *40*(11), 939–941.
24. Arakawa, Y., Sakaki, H., Nishioka, M., Okamoto, H., & Miura, N. (1983). Spontaneous Emission Characteristics of Quantum Well Lasers in Strong Magnetic Fields–An Approach to Quantum-Well-Box Light Source–. *Japanese Journal of Applied Physics*, *22*(12A), L804.
25. Arango, A. C., Oertel, D. C., Xu, Y., Bawendi, M. G., & Bulovic, V. (2009). Heterojunction photovoltaics using printed colloidal quantum dots as a photosensitive layer. *Nano Letters*, *9*(2), 860–863.
26. Arici, E., Sariciftci, N. S., & Meissner, D. (2004). Hybrid solar cells. *Encyclopedia of Nanoscience and Nanotechnology*, *3*, 929–944.
27. Armel, V., Forsyth, M., MacFarlane, D. R., & Pringle, J. M. (2011). Organic ionic plastic crystal electrolytes; a new class of electrolyte for high efficiency solid state dye-sensitized solar cells. *Energy & Environmental Science*, *4*(6), 2234–2239.
28. Arnberg, L., Di Sabatino, M., & Øvrelid, E. (2012). Solidification of Silicon for Solar Cells. *Transactions of the Indian Institute of Metals*, *65*(6), 509–513.
29. Asada, M. (1915). *J Quantum Electron* (1986). *QE-22.*
30. Avrutin, E. A., Marsh, J. H., & Portnoi, E. L. (2000). Monolithic and multi-gigahertz mode-locked semiconductor lasers: constructions, experiments, models and applications. *IEEE Proceedings-Optoelectronics*, *147*(4), 251–278.
31. Bae, W. K., Joo, J., Padilha, L. A., Won, J., Lee, D. C., Lin, Q., & Pietryga, J. M. (2012). Highly effective surface passivation of PbSe quantum dots through reaction with molecular chlorine. *Journal of the American Chemical Society*, *134*(49), 20160–20168.

32. Bae, W. K., Park, Y. S., Lim, J., Lee, D., Padilha, L. A., McDaniel, H., & Klimov, V. I. (2013). Controlling the influence of Auger recombination on the performance of quantum-dot light-emitting diodes. *Nature Communications*, *4*, 2661.

33. Baker, D. R., & Kamat, P. V. (2009). Photosensitization of TiO2 nanostructures with CdS quantum dots: particulate versus tubular support architectures. *Advanced Functional Materials*, *19*(5), 805–811.

34. Baker, J. L., Widmer-Cooper, A., Toney, M. F., Geissler, P. L., & Alivisatos, A. P. (2009). Device-scale perpendicular alignment of colloidal nanorods. *Nano Letters*, *10*(1), 195–201.

35. Baker, S. N., & Baker, G. A. (2010). Luminescent carbon nanodots: emergent nanolights. *Angewandte Chemie International Edition*, *49*(38), 6726–6744.

36. Balberg, I., Savir, E., Jedrzejewski, J., Nassiopoulou, A. G., & Gardelis, S. (2007). Fundamental transport processes in ensembles of silicon quantum dots. *Physical Review B*, *75*(23), 235329.

37. Ballou, B., Lagerholm, B. C., Ernst, L. A., Bruchez, M. P., & Waggoner, A. S. (2004). Noninvasive imaging of quantum dots in mice. *Bioconjugate Chemistry*, *15*(1), 79–86.

38. Bang, B., Koo, H. M., & Park, J. W. (2006, February). Effect of the Working Pressures on Electrical Properties of Al Doped ZnO Thin Films Deposited by dc Magnetron Sputtering. In *Meeting Abstracts* (No. 34, pp. 1340–1340). The Electrochemical Society.

39. Bang, J. H., & Kamat, P. V. (2009). Quantum dot sensitized solar cells. A tale of 2 semiconductor nanocrystals: CdSe and CdTe. *ACS Nano*, *3*(6), 1467–1476.

40. Barkhouse, D. A. R., Debnath, R., Kramer, I. J., Zhitomirsky, D., Pattantyus-Abraham, A. G., Levina, L., & Sargent, E. H. (2011). Depleted bulk heterojunction colloidal quantum dot photovoltaics. *Advanced Materials*, *23*(28), 3134–3138.

41. Bartnik, A. C., Efros, A. L., Koh, W. K., Murray, C. B., & Wise, F. W. (2010). Electronic states and optical properties of PbSe nanorods and nanowires. *Physical Review B*, *82*(19), 195313.

42. Baur, C., Bett, A. W., Dimroth, F., Siefer, G., Meusel, M., Bensch, W., & Strobl, G. (2007). Triple-junction III–V based concentrator solar cells: perspectives and challenges. *Journal of Solar Energy Engineering*, *129*(3), 258–265.

43. Beard, M. C. (2011). Multiple exciton generation in semiconductor quantum dots. *The Journal of Physical Chemistry Letters*, *2*(11), 1282–1288.

44. Beard, M. C., Knutsen, K. P., Yu, P., Luther, J. M., Song, Q., Metzger, W. K., & Nozik, A. J. (2007). Multiple exciton generation in colloidal silicon nanocrystals. *Nano Letters*, *7*(8), 2506–2512.

45. Beard, M. C., Luther, J. M., Semonin, O. E., & Nozik, A. J. (2012). Third generation photovoltaics based on multiple exciton generation in quantum confined semiconductors. *Accounts of Chemical Research*, *46*(6), 1252–1260.

46. Beard, M. C., Midgett, A. G., Hanna, M. C., Luther, J. M., Hughes, B. K., & Nozik, A. J. (2010). Comparing multiple exciton generation in quantum dots to impact ionization in bulk semiconductors: implications for enhancement of solar energy conversion. *Nano Letters*, *10*(8), 3019–3027.

47. Beard, M. C., Midgett, A. G., Law, M., Semonin, O. E., Ellingson, R. J., & Nozik, A. J. (2009). Variations in the quantum efficiency of multiple exciton generation for a series of chemically treated PbSe nanocrystal films .*Nano Letters*, *9*(2), 836–845.

48. Becquerel, E. (1867). *La Lumière, Ses Causes Et Ses Effets* (Vol. 1). Firmin Didot frères, fils et cie.

49. Berberan-Santos, M. N., Bodunov, E. N., & Valeur, B. (2005). Mathematical functions for the analysis of luminescence decays with underlying distributions: 2.Becquerel (compressed hyperbola) and related decay functions. *Chemical Physics*, *317*(1), 57–62.

50. Berg, T. W., Bischoff, S., Magnusdottir, I., & Mork, J. (2001). Ultrafast gain recovery and modulation limitations in self-assembled quantum-dot devices. *IEEE Photonics Technology Letters*, *13*(6), 541–543.

51. Bhunia, S. K., Saha, A., Maity, A. R., Ray, S. C., & Jana, N. R. (2013). Carbon nanoparticle-based fluorescent bioimaging probes. *Scientific Reports*, *3*, 1473.

52. Bimberg, D., & Pohl, U. W. (2011). Quantum dots: promises and accomplishments. *Materials Today*, *14*(9), 388–397.

53. Bimberg, D., Grundmann, M., & Ledentsov, N. N. (1999). *Quantum dot Heterostructures*. John Wiley & Sons.

54. Bisi, O., Ossicini, S., & Pavesi, L. (2000). Porous silicon: a quantum sponge structure for silicon based optoelectronics. *Surface Science*

Reports, *38*(1–3), 1–126.

55. Blackburn, J. L., Selmarten, D. C., Ellingson, R. J., Jones, M., Micic, O., & Nozik, A. J. (2005). Electron and hole transfer from indium phosphide quantum dots. *The Journal of Physical Chemistry B*, *109*(7), 2625–2631.
56. Blow, N. (2009). High-throughput screening: designer screens.
57. Bode, D. E., & Levinstein, H. (1954). Effect of oxygen on the electrical properties of lead telluride films. *Physical Review*, *96*(2), 259.
58. Boercker, J. E., Foos, E. E., Placencia, D., & Tischler, J. G. (2013). Control of PbSe nanorod aspect ratio by limiting phosphine hydrolysis. *Journal of the American Chemical Society*, *135*(40), 15071–15076.
59. Böhm, M. L., Jellicoe, T. C., Rivett, J. P., Sadhanala, A., Davis, N. J., Morgenstern, F. S., & Ehrler, B. (2015). Size and energy level tuning of quantum dot solids via a hybrid ligand complex. *The Journal of Physical Chemistry Letters*, *6*(17), 3510–3514.
60. Böhm, M. L., Jellicoe, T. C., Tabachnyk, M., Davis, N. J., Wisnivesky-Rocca-Rivarola, F., Ducati, C., & Greenham, N. C. (2015). Lead telluride quantum dot solar cells displaying external quantum efficiencies exceeding 120%. *Nano Letters*, *15*(12), 7987–7993.
61. Böhm, M. L., Kist, R. J., Morgenstern, F. S., Ehrler, B., Zarra, S., Kumar, A., & Greenham, N. C. (2014). The influence of nanocrystal aggregates on photovoltaic performance in nanocrystal–polymer bulk heterojunction solar cells. *Advanced Energy Materials*, *4*(12).
62. Borri, P., Langbein, W., Hvam, J. M., Heinrichsdorff, F., Mao, M. H., & Bimberg, D. (2000). Ultrafast gain dynamics in InAs-InGaAs quantum-dot amplifiers. *IEEE Photonics Technology Letters*, *12*(6), 594–596.
63. Borri, P., Schneider, S., Langbein, W., & Bimberg, D. (2006). Ultrafast carrier dynamics in InGaAs quantum dot materials and devices. *Journal of Optics A: Pure and Applied Optics*, *8*(4), S33.
64. Bourlinos, A. B., Stassinopoulos, A., Anglos, D., Zboril, R., Karakassides, M., & Giannelis, E. P. (2008). Surface functionalized carbogenic quantum dots. *Small*, *4*(4), 455–458.
65. Bozyigit, D., Yarema, O., & Wood, V. (2013). Origins of low quantum efficiencies in quantum dot LEDs. *Advanced Functional Materials*, *23*(24), 3024–3029.
66. Brabec, C., Scherf, U., & Dyakonov, V. (Eds.). (2011). *Organic Photovoltaics: Materials, Device Physics, And Manufacturing*

Technologies. John Wiley & Sons.

67. Brendel, R., Werner, J. H., & Queisser, H. J. (1996). Thermodynamic efficiency limits for semiconductor solar cells with carrier multiplication. *Solar Energy Materials and Solar Cells*, *41*, 419–425.

68. Briscoe, J., Marinovic, A., Sevilla, M., Dunn, S., & Titirici, M. (2015). Biomass-derived carbon quantum dot sensitizers for solid-state nanostructured solar cells. *Angewandte Chemie International Edition*, *54*(15), 4463–4468.

69. Brongersma, M. L., Kik, P. G., Polman, A., Min, K. S., & Atwater, H. A. (2000). Size-dependent electron-hole exchange interaction in Si nanocrystals. *Applied Physics Letters*, *76*(3), 351–353.

70. Brown, P. R., Kim, D., Lunt, R. R., Zhao, N., Bawendi, M. G., Grossman, J. C., & Bulović, V. (2014). Energy level modification in lead sulfide quantum dot thin films through ligand exchange. *ACS Nano*, *8*(6), 5863–5872.

71. Bruchez, M., Moronne, M., Gin, P., Weiss, S., & Alivisatos, A. P. (1998). Semiconductor nanocrystals as fluorescent biological labels. *Science*, *281*(5385), 2013–2016.Chan, W. C., & Nie, S. (1998). Quantum dot bioconjugates for ultrasensitive nonisotopic detection. *Science*, *281*(5385), 2016–2018.

72. Brus, L. E. (1984). Electron–electron and electron-hole interactions in small semiconductor crystallites: The size dependence of the lowest excited electronic state. *The Journal of Chemical Physics*, *80*(9), 4403–4409.

73. Bude, J., & Hess, K. (1992). Thresholds of impact ionization in semiconductors. *Journal of Applied Physics*, *72*(8), 3554–3561.

74. Buhbut, S., Itzhakov, S., Tauber, E., Shalom, M., Hod, I., Geiger, T., & Zaban, A. (2010). Built-in quantum dot antennas in dye-sensitized solar cells. *ACS Nano*, *4*(3), 1293–1298.

75. Burschka, J., Pellet, N., Moon, S. J., Humphry-Baker, R., Gao, P., Nazeeruddin, M. K., & Grätzel, M. (2013). Sequential deposition as a route to high-performance perovskite-sensitized solar cells. *Nature*, *499*(7458), 316.

76. Cai, X., Tan, S., Lin, M., Xie, A., Mai, W., Zhang, X., & Liu, Y. (2011). Synergistic antibacterial brilliant blue/reduced graphene oxide/quaternary phosphonium salt composite with excellent water solubility and specific targeting capability. *Langmuir*, *27*(12), 7828–7835.

77. Cai, X., Tan, S., Yu, A., Zhang, J., Liu, J., Mai, W., & Jiang, Z. (2012). Sodium 1-Naphthalenesulfonate-Functionalized Reduced Graphene Oxide Stabilizes Silver Nanoparticles with Lower Cytotoxicity and Long-Term Antibacterial Activity. *Chemistry–An Asian Journal*, *7*(7), 1664–1670.

78. Canham, L. T. (1990). Silicon quantum wire array fabrication by electrochemical and chemical dissolution of wafers. *Applied Physics Letters*, *57*(10), 1046–1048.

79. Cansizoglu, H., Cansizoglu, M. F., Finckenor, M., & Karabacak, T. (2013). Optical absorption properties of semiconducting nanostructures with different shapes. *Advanced Optical Materials*, *1*(2), 158–166.

80. Cao, A., Liu, Z., Chu, S., Wu, M., Ye, Z., Cai, Z., & Liu, Y. (2010). A Facile One-step Method to Produce Graphene–CdS Quantum Dot Nanocomposites as Promising Optoelectronic Materials. *Advanced Materials*, *22*(1), 103–106.

81. Cao, L., Wang, X., Meziani, M. J., Lu, F., Wang, H., Luo, P. G., & Xie, S. Y. (2007). Carbon dots for multiphoton bioimaging. *Journal of the American Chemical Society*, *129*(37), 11318–11319.

82. Carey, G. H., Abdelhady, A. L., Ning, Z., Thon, S. M., Bakr, O. M., & Sargent, E. H. (2015). Colloidal quantum dot solar cells. *Chemical Reviews*, *115*(23), 12732–12763.

83. Caruge, J. M., Halpert, J. E., Wood, V., Bulović, V., & Bawendi, M. G. (2008). Colloidal quantum-dot light-emitting diodes with metal-oxide charge transport layers. *Nature Photonics*, *2*(4), 247.

84. Chang, C. H., & Lee, Y. L. (2007). Chemical bath deposition of CdS quantum dots onto mesoscopic Ti O 2 films for application in quantum-dot-sensitized solar cells. *Applied Physics Letters, 91*(5), 053503.

85. Chang, H., Lv, X., Zhang, H., & Li, J. (2010). Quantum dots sensitized graphene: In situ growth and application in photoelectrochemical cells. *Electrochemistry Communications*, *12*(3), 483–487.

86. Chang, J. A., Rhee, J. H., Im, S. H., Lee, Y. H., Kim, H. J., Seok, S. I., ... & Gratzel, M. (2010). High-performance nanostructured inorganic−organic heterojunction solar cells. *Nano Letters*, *10*(7), 2609–2612.

87. Chaoui, R., Mahmoudi, B., & Si Ahmed, Y. (2008). Porous silicon antireflection layer for solar cells using metal-assisted chemical etching. *Physica Status Solidi (a)*, *205*(7), 1724–1728.

88. Chapin, D. M., Fuller, C. S., & Pearson, G. L. (1954). A new silicon p-n

junction photocell for converting solar radiation into electrical power. *Journal of Applied Physics*, *25*(5), 676–677.

89. Chappell, H. E., Hughes, B. K., Beard, M. C., Nozik, A. J., & Johnson, J. C. (2011). Emission quenching in PbSe quantum dot arrays by short-term air exposure. *The Journal of Physical Chemistry Letters*, *2*(8), 889–893.

90. Chelikowsky, J. R., & Cohen, M. L. (1976). Nonlocal pseudopotential calculations for the electronic structure of eleven diamond and zinc-blende semiconductors. *Physical Review B*, *14*(2), 556.

91. Chen, H. Y., Lo, M. K., Yang, G., Monbouquette, H. G., & Yang, Y. (2008). Nanoparticle-assisted high photoconductive gain in composites of polymer and fullerene. *Nature Nanotechnology*, *3*(9), 543.

92. Chen, J., Lei, W., & Deng, W. Q. (2011). Reduced charge recombination in a co-sensitized quantum dot solar cell with 2 different sizes of CdSe quantum dot. *Nanoscale*, *3*(2), 674–677.

93. Chen, J., Li, C., Zhao, D. W., Lei, W., Zhang, Y., Cole, M. T., ... & Milne, W. I. (2010). A quantum dot sensitized solar cell based on vertically aligned carbon nanotube templated ZnO arrays. *Electrochemistry Communications*, *12*(10), 1432–1435.

94. Chen, J., Pan, J., Huang, Q., Xu, F., Zhang, Z., Lei, W., & Nathan, A. (2015). Graphene oxide/PEDOT: PSS as injection layer for quantum dot light emitting diode. *Physica Status Solidi (a)*, *212*(12), 2856–2861.

95. Chen, J., Song, J. L., Sun, X. W., Deng, W. Q., Jiang, C. Y., Lei, W., & Liu, R. S. (2009). An oleic acid-capped CdSe quantum-dot sensitized solar cell. *Applied Physics Letters*, *94*(15), 153115.

96. Chen, J., Wu, J., Lei, W., Song, J. L., Deng, W. Q., & Sun, X. W. (2010). Co-sensitized quantum dot solar cell based on ZnO nanowire. *Applied Surface Science*, *256*(24), 7438–7441.

97. Chen, J., Zhao, D. W., Song, J. L., Sun, X. W., Deng, W. Q., Liu, X. W., & Lei, W. (2009). Directly assembled CdSe quantum dots on TiO2 in aqueous solution by adjusting pH value for quantum dot sensitized solar cells. *Electrochemistry Communications*, *11*(12), 2265–2267.

98. Chen, O., Wei, H., Maurice, A., Bawendi, M., & Reiss, P. (2013). Pure colors from core–shell quantum dots. *MRS Bulletin*, *38*(9), 696–702.

99. Chen, O., Zhao, J., Chauhan, V. P., Cui, J., Wong, C., Harris, D. K., & Bawendi, M. G. (2013). Compact high-quality CdSe–CdS core–shell nanocrystals with narrow emission linewidths and suppressed blinking.

Nature Materials, *12*(5), 445.

100. Chen, S., Paulose, M., Ruan, C., Mor, G. K., Varghese, O. K., Kouzoudis, D., & Grimes, C. A. (2006). Electrochemically synthesized CdS nanoparticle-modified TiO2 nanotube-array photoelectrodes: preparation, characterization, and application to photoelectrochemical cells. *Journal of Photochemistry and Photobiology A: Chemistry*, *177*(2–3), 177–184.
101. Chen, Y., Vela, J., Htoon, H., Casson, J. L., Werder, D. J., Bussian, D. A., & Hollingsworth, J. A. (2008). "Giant" multishell CdSe nanocrystal quantum dots with suppressed blinking. *Journal of the American Chemical Society*, *130*(15), 5026–5027.
102. Chen, Z., Berciaud, S., Nuckolls, C., Heinz, T. F., & Brus, L. E. (2010). Energy transfer from individual semiconductor nanocrystals to graphene. *ACS Nano*, *4*(5), 2964–2968.
103. Chiba, Y., Islam, A., Watanabe, Y., Komiya, R., Koide, N., & Han, L. (2006). Dye-sensitized solar cells with conversion efficiency of 11.1%. *Japanese Journal of Applied Physics*, *45*(7L), L638.
104. Cho, K. S., Lee, E. K., Joo, W. J., Jang, E., Kim, T. H., Lee, S. J., & Kim, J. M. (2009). High-performance crosslinked colloidal quantum-dot light-emitting diodes. *Nature Photonics*, *3*(6), 341.
105. Choi, J. J., Lim, Y. F., Santiago-Berrios, M. E. B., Oh, M., Hyun, B. R., Sun, L., & Wise, F. W. (2009). PbSe nanocrystal excitonic solar cells. *Nano Letters*, *9*(11), 3749–3755.
106. Choi, J. J., Wenger, W. N., Hoffman, R. S., Lim, Y. F., Luria, J., Jasieniak, J., & Hanrath, T. (2011). Solution-Processed Nanocrystal Quantum Dot Tandem Solar Cells. *Advanced Materials*, *23*(28), 3144–3148.
107. Christensen, O. (1976). Quantum efficiency of the internal photoelectric effect in silicon and germanium. *Journal of Applied Physics*, *47*(2), 689–695.
108. Chuah, L. S., Hassan, Z., & Hassan, H. A. (2010, July). Visible-Light Emission Due to Quantum Size Effects in Porous Crystalline Silicon. In *AIP Conference Proceedings* (*Vol. 1250*, No. 1, pp. 385–388). AIP.
109. Chuang, C. H. M., Brown, P. R., Bulović, V., & Bawendi, M. G. (2014). Improved performance and stability in quantum dot solar cells through band alignment engineering. *Nature Materials*, *13*(8), 796.
110. Chuang, C. H. M., Maurano, A., Brandt, R. E., Hwang, G. W., Jean,

J., Buonassisi, T., ... & Bawendi, M. G. (2015). Open-circuit voltage deficit, radiative sub-bandgap states, and prospects in quantum dot solar cells. *Nano letters*, *15*(5), 3286–3294.

111. Chung, J. H., Ah, C. S., & Jang, D. J. (2001). Formation and distinctive decay times of surface-and lattice-bound Mn2+ impurity luminescence in ZnS nanoparticles. *The Journal of Physical Chemistry B*, *105*(19), 4128–4132.
112. Church, C. P., Muthuswamy, E., Zhai, G., Kauzlarich, S. M., & Carter, S. A. (2013). Quantum dot Ge/TiO_2 heterojunction photoconductor fabrication and performance. *Applied Physics Letters*, *103*(22), 223506.
113. Cirloganu, C. M., Padilha, L. A., Lin, Q., Makarov, N. S., Velizhanin, K. A., Luo, H., & Klimov, V. I. (2014). Enhanced carrier multiplication in engineered quasi-type-II quantum dots. *Nature Communications*, *5*, 4148.
114. Clifford, J. P., Johnston, K. W., Levina, L., & Sargent, E. H. (2007). Schottky barriers to colloidal quantum dot films. *Applied Physics Letters*, *91*(25), 253117.
115. Cloke, R. R. (2015). Synthesis of doped graphene nanoribbons from molecular and polymeric precursors. University of California, Berkeley.
116. Coe-Sullivan, S., Steckel, J. S., Woo, W. K., Bawendi, M. G., & Bulović, V. (2005). Large-Area Ordered Quantum-Dot Monolayers via Phase Separation During Spin-Casting. *Advanced Functional Materials*, *15*(7), 1117–1124.
117. Colvin, V. L., Schlamp, M. C., & Alivisatos, A. P. (1994). Light-emitting diodes made from cadmium selenide nanocrystals and a semiconducting polymer. *Nature*, *370*(6488), 354.
118. Congreve, D. N., Lee, J., Thompson, N. J., Hontz, E., Yost, S. R., Reusswig, P. D., & Baldo, M. A. (2013). External quantum efficiency above 100% in a singlet-exciton-fission–based organic photovoltaic cell. *Science*, *340*(6130), 334–337.
119. Conibeer, G. J., Jiang, C. W., König, D., Shrestha, S., Walsh, T., & Green, M. A. (2008). Selective energy contacts for hot carrier solar cells. *Thin Solid Films*, *516*(20), 6968–6973.
120. Conibeer, G., Green, M., Cho, E. C., König, D., Cho, Y. H., Fangsuwannarak, T., & Huang, S. (2008). Silicon quantum dot nanostructures for tandem photovoltaic cells. *Thin Solid Films*, *516*(20), 6748–6756.

121. Conibeer, G., Green, M., Corkish, R., Cho, Y., Cho, E. C., Jiang, C. W., & Trupke, T. (2006). Silicon nanostructures for third generation photovoltaic solar cells. *Thin Solid Films*, *511*, 654–662.

122. Contestabile, G., Maruta, A., Sekiguchi, S., Morito, K., Sugawara, M., & Kitayama, K. I. (2010). Cross-gain modulation in quantum-dot SOA at 1550 nm. *IEEE Journal of Quantum Electronics*, *46*(12), 1696–1703.

123. Cragg, G. E., & Efros, A. L. (2009). Suppression of Auger processes in confined structures. *Nano Letters*, *10*(1), 313–317.

124. Crisp, R. W., Kroupa, D. M., Marshall, A. R., Miller, E. M., Zhang, J., Beard, M. C., & Luther, J. M. (2015). Metal halide solid-state surface treatment for high efficiency PbS and PbSe QD solar cells. *Scientific Reports*, *5*, 9945.

125. Cuesta, I. G., Pownall, B., Pelloni, S., & de Merás, A. M. S. (2015). π-Ring currents in doped coronenes with nitrogen and boron: diatropic–paratropic duality. *Physical Chemistry Chemical Physics*, *17*(38), 24885–24894.

126. Cullis, A. G., & Canham, L. T. (1991). Visible light emission due to quantum size effects in highly porous crystalline silicon. *Nature*, *353*(6342), 335.

127. Cullis, A. G., Canham, L. T., & Calcott, P. D. J. (1997). The structural and luminescence properties of porous silicon. *Journal of Applied Physics*, *82*(3), 909–965.

128. Cunningham, P. D., Boercker, J. E., Foos, E. E., Lumb, M. P., Smith, A. R., Tischler, J. G., & Melinger, J. S. (2011). Enhanced multiple exciton generation in quasi-one-dimensional semiconductors. *Nano Letters*, *11*(8), 3476–3481.

129. Dabbousi, B. O., Rodriguez-Viejo, J., Mikulec, F. V., Heine, J. R., Mattoussi, H., Ober, R., & Bawendi, M. G. (1997). (CdSe) ZnS core-shell quantum dots: synthesis and characterization of a size series of highly luminescent nanocrystallites. *The Journal of Physical Chemistry B*, *101*(46), 9463–9475.

130. Dai, L. (2010). Layered graphene/quantum dots: nanoassemblies for highly efficient solar cells. *ChemSusChem*, *3*(7), 797–799.

131. Dai, X., Zhang, Z., Jin, Y., Niu, Y., Cao, H., Liang, X., & Peng, X. (2014). Solution-processed, high-performance light-emitting diodes based on quantum dots. *Nature*, *515*(7525), 96.

132. Daldosso, N., Luppi, M., Ossicini, S., Degoli, E., Magri, R., Dalba, G.,

& Boninelli, S. (2003). Role of the interface region on the optoelectronic properties of silicon nanocrystals embedded in SiO_2. *Physical Review B*, *68*(8), 085327.

133. Dalven, R. (1971). Energy-gap anomaly in the semiconductor sequence PbS, PbSe, and PbTe. *Physical Review B*, *3*(10), 3359.
134. Damtie, F. A., Karki, K. J., Pullerits, T., & Wacker, A. (2016). Optimization schemes for efficient multiple exciton generation and extraction in colloidal quantum dots. *The Journal of Chemical Physics*, *145*(6), 064703.
135. Darakchieva, V., Baleva, M., Surtchev, M., & Goranova, E. (2000). Structural and optical analysis of β− FeSi 2 thin layers prepared by ion-beam synthesis and solid-state reaction. *Physical Review B*, *62*(19), 13057.
136. Davis, N. J., Böhm, M. L., Tabachnyk, M., Wisnivesky-Rocca-Rivarola, F., Jellicoe, T. C., Ducati, C., & Greenham, N. C. (2015). Multiple-exciton generation in lead selenide nanorod solar cells with external quantum efficiencies exceeding 120%. *Nature communications*, *6*, 8259.
137. Dayal, S., Kopidakis, N., Olson, D. C., Ginley, D. S., & Rumbles, G. (2009). Photovoltaic devices with a low band gap polymer and CdSe nanostructures exceeding 3% efficiency. *Nano Letters*, *10*(1), 239–242.
138. De Geyter, B., Justo, Y., Moreels, I., Lambert, K., Smet, P. F., Van Thourhout, D., & Vanmaekelbergh, D. (2010). The different nature of band edge absorption and emission in colloidal PbSe/CdSe core/shell quantum dots. *ACS Nano*, *5*(1), 58–66.
139. De Vos, A. (1980). Detailed balance limit of the efficiency of tandem solar cells. *Journal of Physics D: Applied Physics*, *13*(5), 839.
140. De, B., & Karak, N. (2013). A green and facile approach for the synthesis of water soluble fluorescent carbon dots from banana juice. Rsc Advances, 3(22), 8286–8290.
141. Del Canizo, C., Del Coso, G., & Sinke, W. C. (2009). Crystalline silicon solar module technology: Towards the 1€ per watt-peak goal. *Progress in Photovoltaics: Research and Applications*, *17*(3), 199–209.
142. Di, D., Perez-Wurfl, I., Conibeer, G., & Green, M. A. (2010). Formation and photoluminescence of Si quantum dots in SiO2/Si3N4 hybrid matrix for all-Si tandem solar cells. *Solar Energy Materials and Solar Cells*, *94*(12), 2238–2243.

143. Dietmueller, R., Stegner, A. R., Lechner, R., Niesar, S., Pereira, R. N., Brandt, M. S., & Stutzmann, M. (2009). Light-induced charge transfer in hybrid composites of organic semiconductors and silicon nanocrystals. *Applied Physics Letters*, *94*(11), 83.

144. DiMaria, D. J., Dong, D. W., Pesavento, F. L., Lam, C., & Brorson, S. D. (1984). Enhanced conduction and minimized charge trapping in electrically alterable read-only memories using off-stoichiometric silicon dioxide films. *Journal of Applied Physics*, *55*(8), 3000–3019.

145. DiMaria, D. J., Kirtley, J. R., Pakulis, E. J., Dong, D. W., Kuan, T. S., Pesavento, F. L., Brorson, S. D., et al. (1984). Electroluminescence studies in silicon dioxide films containing tiny silicon islands. *Journal of Applied Physics*, *56*(2), 401–416.

146. Dimroth, F., Grave, M., Beutel, P., Fiedeler, U., Karcher, C., Tibbits, T. N., & Bett, A. W. (2014). Wafer bonded four-junction GaInP/GaAs//GaInAsP/GaInAs concentrator solar cells with 44.7% efficiency. *Progress in Photovoltaics: Research and Applications*, *22*(3), 277–282.

147. Ding, C., Zhu, A., & Tian, Y. (2013). Functional surface engineering of C-dots for fluorescent biosensing and in vivo bioimaging. *Accounts of Chemical Research*, *47*(1), 20–30.

148. Dingle, R., & Henry, C. H. (1976). U.S. Patent No. 3,982,207. Washington, DC: U.S. Patent and Trademark Office.

149. Doany, F. E., & Grischkowsky, D. (1988). Measurement of ultrafast hot-carrier relaxation in silicon by thin-film-enhanced, time-resolved reflectivity. *Applied Physics Letters*, *52*(1), 36–38.

150. Dong, Y., Shao, J., Chen, C., Li, H., Wang, R., Chi, Y., & Chen, G. (2012). Blue luminescent graphene quantum dots and graphene oxide prepared by tuning the carbonization degree of citric acid. *Carbon*, *50*(12), 4738–4743.

151. Dovrat, M., Goshen, Y., Jedrzejewski, J., Balberg, I., & Sa'Ar, A. (2004). Radiative versus nonradiative decay processes in silicon nanocrystals probed by time-resolved photoluminescence spectroscopy. *Physical Review B*, *69*(15), 155311.

152. Du, J., Du, Z., Hu, J. S., Pan, Z., Shen, Q., Sun, J., & Wan, L. J. (2016). Zn–Cu–In–Se quantum dot solar cells with a certified power conversion efficiency of 11.6%. *Journal of the American Chemical Society*, *138*(12), 4201–4209.

153. Dualeh, A., Moehl, T., Tétreault, N., Teuscher, J., Gao, P., Nazeeruddin,

M. K., & Grätzel, M. (2013). Impedance spectroscopic analysis of lead iodide perovskite-sensitized solid-state solar cells. *ACS Nano*, *8*(1), 362–373.

154. Dutta, A. (2000). Hayafune Y. Oda S. Single electron memory devices based on plasm a—derived silicon nanocrystals. *J 户 d Appl P*, *39*(8B), L855.

155. Edmonds, J. A., Wise, M. A., & MacCracken, C. N. (1994). Advanced energy technologies and climate change: An analysis using the global change assessment model (GCAM) (No. PNL-9798). Pacific Northwest National Laboratory (PNNL), Richland, WA (United States).

156. Efros, A. L., & Rosen, M. (1997). Random telegraph signal in the photoluminescence intensity of a single quantum dot. *Physical Review Letters*, *78*(6), 1110.

157. Ehrlich, H. C., & Rarey, M. (2011). Maximum common subgraph isomorphism algorithms and their applications in molecular science: a review. *Wiley Interdisciplinary Reviews: Computational Molecular Science*, *1*(1), 68–79.

158. El-Den, M. B., & El-Nahass, M. M. (2003). Optical properties of AsSe1.5− xTex glassy system. *Optics & Laser Technology*, *35*(5), 335–340.

159. El-Nahass, M. M., & El Den, M. B. (2001). Dispersive optical constants of glassy AsSe1− xTex system. *Optics & Laser Technology*, *33*(1), 31–35.

160. Ellingson, R. J., Beard, M. C., Johnson, J. C., Yu, P., Micic, O. I., Nozik, A. J., & Efros, A. L. (2005). Highly efficient multiple exciton generation in colloidal PbSe and PbS quantum dots. *Nano letters*, *5*(5), 865–871.

161. Emery, K. (2011). Measurement and characterization of solar cells and modules. *Handbook of Photovoltaic Science and Engineering,* Second Edition, 797–840.

162. Emin, S. M., Sogoshi, N., Nakabayashi, S., Fujihara, T., & Dushkin, C. D. (2009). Kinetics of photochromic induced energy transfer between manganese-doped zinc-selenide quantum dots and spiropyrans. *The Journal of Physical Chemistry C*, *113*(10), 3998–4007.

163. Emin, S., Loukanov, A., Wakasa, M., Nakabayashi, S., & Kaneko, Y. (2010). Photostability of water-dispersible CdTe quantum dots: capping ligands and oxygen. *Chemistry Letters*, *39*(6), 654–656.

164. Emin, S., Sogoshi, N., Nakabayashi, S., Villeneuve, M., & Dushkin, C. (2009). Growth kinetics of CdS quantum dots and synthesis of their polymer nano-composites in CTAB reverse micelles. *Journal of Photochemistry and Photobiology A: Chemistry*, *207*(2–3), 173–180.

165. Empedocles, S. A., & Bawendi, M. G. (1997). Quantum-confined stark effect in single CdSe nanocrystallite quantum dots. *Science*, *278*(5346), 2114–2117.

166. Empedocles, S. A., Norris, D. J., & Bawendi, M. G. (1996). Photoluminescence spectroscopy of single CdSe nanocrystallite quantum dots. *Physical Review Letters*, *77*(18), 3873.

167. Erslev, P. T., Chen, H. Y., Gao, J., Beard, M. C., Frank, A. J., van de Lagemaat, J., & Luther, J. M. (2012). Sharp exponential band tails in highly disordered lead sulfide quantum dot arrays. *Physical Review B*, *86*(15), 155313.

168. Etgar, L. (2013). Semiconductor nanocrystals as light harvesters in solar cells. *Materials*, *6*(2), 445–459.

169. Etgar, L., Gao, P., Xue, Z., Peng, Q., Chandiran, A. K., Liu, B., & Grätzel, M. (2012). Mesoscopic CH3NH3PbI3/TiO2 heterojunction solar cells. *Journal of the American Chemical Society*, *134*(42), 17396–17399.

170. Fadel, M., Omer, O. A., & Basily, R. R. (1998). A study of some optical properties of hafnium dioxide (HfO2) thin films and their applications. *Applied Physics A*, *66*(3), 335–343.

171. Feltrin, A., & Freundlich, A. (2008). Material considerations for terawatt level deployment of photovoltaics. *Renewable Energy*, *33*(2), 180–185.

172. Feng, Z. C., & Tsu, R. (1994). *Porous Silicon*. World Scientific.

173. Fernandez, B. G., Lopez, M., Garcia, C., Perez-Rodriguez, A., Morante, J. R., Bonafos, C., & Claverie, A. (2002). Influence of average size and interface passivation on the spectral emission of Si nanocrystals embedded in SiO~ 2.*Journal of Applied Physics*, *91*(2), 798–807.

174. Forrest, S. R., Kaplan, M. L., Schmidt, P. H., Feldmann, W. L., & Yanowski, E. (1982). Organic-on-inorganic semiconductor contact barrier devices. *Applied Physics Letters*, *41*(1), 90–93.

175. Frankl, P., Nowak, S., Gutschner, M., Gnos, S., & Rinke, T. (2010). Technology roadmap: solar photovoltaic energy. *International Energy Association*.

176. Franzo, G., Irrera, A., Moreira, E. C., Miritello, M., Iacona, F., Sanfilippo, D., & Priolo, F. (2002). Electroluminescence of silicon nanocrystals in MOS structures. *Applied Physics A*, *74*(1), 1–5.

177. Fritzinger, B., Capek, R. K., Lambert, K., Martins, J. C., & Hens, Z. (2010). Utilizing self-exchange to address the binding of carboxylic acid ligands to CdSe quantum dots. *Journal of the American Chemical Society*, *132*(29), 10195–10201.

178. Fu, K., Yue, Q., Gao, B., Sun, Y. and Zhu, L. (2013). Preparation, characterization and application of lignin-based activated carbon from black liquor lignin by steam activation. *Chemical Engineering Journal, 228*,1074–1082.

179. Fujishima, A., & Honda, K. (1972). Electrochemical photolysis of water at a semiconductor electrode.*Nature*, *238*(5358), 37.

180. Galland, C., Ghosh, Y., Steinbrück, A., Sykora, M., Hollingsworth, J. A., Klimov, V. I., & Htoon, H. (2011). Two types of luminescence blinking revealed by spectroelectrochemistry of single quantum dots. *Nature*, *479*(7372), 203.

181. Gao, B., Shen, C., Zhang, B., Zhang, M., Yuan, S., Yang, Y., & Chen, G. (2014). Green synthesis of highly efficient CdSe quantum dots for quantum-dots-sensitized solar cells. *Journal of Applied Physics*, *115*(19), 193104.

182. Gao, J., & Johnson, J. C. (2012). Charge trapping in bright and dark states of coupled PbS quantum dot films. *ACS Nano*, *6*(4), 3292–3303.

183. Gao, J., Jeong, S., Lin, F., Erslev, P. T., Semonin, O. E., Luther, J. M., & Beard, M. C. (2013). Improvement in carrier transport properties by mild thermal annealing of PbS quantum dot solar cells. *Applied Physics Letters*, *102*(4), 043506.

184. Gao, J., Luther, J. M., Semonin, O. E., Ellingson, R. J., Nozik, A. J., & Beard, M. C. (2011). Quantum dot size dependent J– V characteristics in heterojunction ZnO/PbS quantum dot solar cells. *Nano letters*, *11*(3), 1002–1008.

185. Gao, J., Perkins, C. L., Luther, J. M., Hanna, M. C., Chen, H. Y., Semonin, O. E., & Beard, M. C. (2011). n-Type transition metal oxide as a hole extraction layer in PbS quantum dot solar cells. *Nano Letters*, *11*(8), 3263–3266.

186. Gao, X., Cui, Y., Levenson, R. M., Chung, L. W., & Nie, S. (2004). In vivo cancer targeting and imaging with semiconductor quantum dots.

Nature Biotechnology, *22*(8), 969.

187. Gao, Y., Aerts, M., Sandeep, C. S., Talgorn, E., Savenije, T. J., Kinge, S., & Houtepen, A. J. (2012). Photoconductivity of PbSe quantum-dot solids: dependence on ligand anchor group and length. *ACS Nano*, *6*(11), 9606–9614.

188. Gap, B. (2010). Composition Engineering on a Nanocrystal in Solution Peng, Xiaogang. *Accounts of Chemical Research*, *43*(11), 1387–1395.

189. García-Santamaría, F., Brovelli, S., Viswanatha, R., Hollingsworth, J. A., Htoon, H., Crooker, S. A., & Klimov, V. I. (2011). Breakdown of volume scaling in Auger recombination in CdSe/CdS heteronanocrystals: the role of the core– shell interface. *Nano Letters*, *11*(2), 687–693.

190. García-Santamaría, F., Chen, Y., Vela, J., Schaller, R. D., Hollingsworth, J. A., & Klimov, V. I. (2009). Suppressed auger recombination in "giant" nanocrystals boosts optical gain performance. *Nano Letters*, *9*(10), 3482–3488.

191. Gaudreau, L., Tielrooij, K. J., Prawiroatmodjo, G. E. D. K., Osmond, J., de Abajo, F. G., & Koppens, F. H. L. (2013). Universal distance-scaling of nonradiative energy transfer to graphene. *Nano Letters*, *13*(5), 2030–2035.

192. Ge, J., Lan, M., Zhou, B., Liu, W., Guo, L., Wang, H., & Meng, X. (2014). A graphene quantum dot photodynamic therapy agent with high singlet oxygen generation. *Nature Communications*, *5*, 4596.

193. Gehrig, E., Hess, O., Ribbat, C., Sellin, R. L., & Bimberg, D. (2004). Dynamic filamentation and beam quality of quantum-dot lasers. *Applied Physics Letters*, *84*(10), 1650–1652.

194. Geim, A. K., & Novoselov, K. S. (2007). The rise of graphene. *Nature Materials*, *6*(3), 183.

195. Geiregat, P., Delerue, C., Justo, Y., Aerts, M., Spoor, F., Van Thourhout, D., & Hens, Z. (2015). A phonon scattering bottleneck for carrier cooling in lead chalcogenide nanocrystals. *ACS Nano*, *9*(1), 778–788.

196. Geng, X., Niu, L., Xing, Z., Song, R., Liu, G., Sun, M., & Sun, L. (2010). Aqueous-Processable Noncovalent Chemically Converted Graphene–Quantum Dot Composites for Flexible and Transparent Optoelectronic Films. *Advanced Materials*, *22*(5), 638–642.

197. Genovese, M. P., Lightcap, I. V., & Kamat, P. V. (2011). Sun-believable solar paint. A transformative one-step approach for designing nanocrystalline solar cells. *ACS Nano*, *6*(1), 865–872.

198. Germann, T. D., Hofmann, W., Nadtochiy, A. M., Schulze, J. H., Mutig, A., Strittmatter, A., & Bimberg, D. (2012). Electro-optical resonance modulation of vertical-cavity surface-emitting lasers. *Optics Express*, *20*(5), 5099–5107.

199. Germann, T. D., Strittmatter, A., Mutig, A., Nadtochiy, A. M., Lott, J. A., Blokhin, S. A., & Bimberg, D. (2010). Monolithic electro-optically modulated vertical cavity surface emitting laser with 10 Gb/s open-eye operation. *Physica Status Solidi (c)*, *7*(10), 2552–2554.

200. Ghazinejad, M., Guo, S., Wang, W., Ozkan, M., & Ozkan, C. S. (2013). Synchronous chemical vapor deposition of large-area hybrid graphene–carbon nanotube architectures. *Journal of Materials Research*, *28*(7), 958–968.

201. Gimenez, S., Mora-Sero, I., Macor, L., Guijarro, N., Lana-Villarreal, T., Gomez, R., & Bisquert, J. (2009). Improving the performance of colloidal quantum-dot-sensitized solar cells. *Nanotechnology*, *20*(29), 295204.

202. Ginger, D. S., & Greenham, N. C. (1999). Photoinduced electron transfer from conjugated polymers to CdSe nanocrystals. *Physical Review B*, *59*(16), 10622.

203. Ginger, D. S., & Greenham, N. C. (2000). Charge injection and transport in films of CdSe nanocrystals. *Journal of Applied Physics*, *87*(3), 1361–1368.

204. Gloeckler, M., & Sites, J. R. (2004). Apparent quantum efficiency effects in CdTe solar cells. *Journal of Applied Physics*, *95*(8), 4438–4445.

205. Godefroo, S., Hayne, M., Jivanescu, M., Stesmans, A., Zacharias, M., Lebedev, O. I., & Moshchalkov, V. V. (2008). Classification and control of the origin of photoluminescence from Si nanocrystals. *Nature Nanotechnology*, *3*(3), 174.

206. Gong, X., Yang, Z., Walters, G., Comin, R., Ning, Z., Beauregard, E., & Sargent, E. H. (2016). Highly efficient quantum dot near-infrared light-emitting diodes. *Nature Photonics*, *10*(4), 253.

207. Gonzalez-Pedro, V., Juarez-Perez, E. J., Arsyad, W. S., Barea, E. M., Fabregat-Santiago, F., Mora-Sero, I., & Bisquert, J. (2014). General working principles of CH3NH3PbX3 perovskite solar cells. *Nano Letters*, *14*(2), 888–893.

208. González-Pedro, V., Xu, X., Mora-Sero, I., & Bisquert, J. (2010).

Modeling high-efficiency quantum dot sensitized solar cells. *ACS Nano*, *4*(10), 5783–5790.

209. Goodwin, E. D., Diroll, B. T., Oh, S. J., Paik, T., Murray, C. B., & Kagan, C. R. (2014). Effects of post-synthesis processing on CdSe nanocrystals and their solids: Correlation between surface chemistry and optoelectronic properties. *The Journal of Physical Chemistry C*, *118*(46), 27097–27105.
210. Goodwin, E. D., Straus, D. B., Gaulding, E. A., Murray, C. B., & Kagan, C. R. (2016). The effects of inorganic surface treatments on photogenerated carrier mobility and lifetime in PbSe quantum dot thin films. *Chemical Physics*, *471*, 81–88.
211. Gowrishankar, V., Scully, S. R., McGehee, M. D., Wang, Q., & Branz, H. M. (2006). Exciton splitting and carrier transport across the amorphous-silicon/polymer solar cell interface. *Applied Physics Letters*, *89*(25), 252102.
212. Graetzel, M., Janssen, R. A., Mitzi, D. B., & Sargent, E. H. (2012). Materials interface engineering for solution-processed photovoltaics. *Nature*, *488*(7411), 304.
213. Graphene, L., & Dots, Q. (2010). Nanoassemblies for Highly Efficient Solar Cells Dai, Liming. *ChemSusChem*, *3*(7), 797–799.
214. Grätzel, M. (2001). Photoelectrochemical cells. *Nature*, *414*(6861), 338.
215. Green, M. A. (2003). Third generation photovoltaics: advanced solar electricity generation. *Springer-Verlag, Berlin*, 4.
216. Green, M. A., Emery, K., Hishikawa, Y., & Warta, W. (2010). Solar cell efficiency tables (version 36). *Progress in Photovoltaics: Research and Applications*, *18*(5), 346–352.
217. Green, M. A., Emery, K., Hishikawa, Y., Warta, W., & Dunlop, E. D. (2015). Solar cell efficiency tables (Version 45). *Progress in Photovoltaics: Research and Applications*, *23*(1), 1–9.
218. Greenham, N. C., Peng, X., & Alivisatos, A. P. (1996). Charge separation and transport in conjugated-polymer/semiconductor-nanocrystal composites studied by photoluminescence quenching and photoconductivity. *Physical Review B*, *54*(24), 17628.
219. Grundmann, M. (1999). The present status of quantum dot lasers. *Physica E: Low-dimensional Systems and Nanostructures*, *5*(3), 167–184.

220. Grundmann, M., Heinrichsdorff, F., Ledentsov, N. N., Bimberg, D., & Alferov, Z. I. (1999). Diode lasers based on quantum dots. In *Advances in Solid State Physics 38* (pp. 203–214). Springer, Berlin, Heidelberg.

221. Grundmann, M., Heinrichsdorff, F., Ribbat, C., Mao, M. H., & Bimberg, D. (1999). Quantum dot lasers: recent progress in theoretical understanding and demonstration of high-output-power operation. *Applied Physics B*, *69*(5–6), 413–416.

222. Guldi, D. M., Rahman, G. A., Sgobba, V., Kotov, N. A., Bonifazi, D., & Prato, M. (2006). CNT− CdTe versatile donor− acceptor nanohybrids. *Journal of the American Chemical Society*, *128*(7), 2315–2323.

223. Günes, S., Fritz, K. P., Neugebauer, H., Sariciftci, N. S., Kumar, S., & Scholes, G. D. (2007). Hybrid solar cells using PbS nanoparticles. *Solar Energy Materials and Solar Cells*, *91*(5), 420–423.

224. Günes, S., Neugebauer, H., & Sariciftci, N. S. (2007). Conjugated polymer-based organic solar cells. *Chemical Reviews*, *107*(4), 1324–1338.

225. Guo, C. X., Yang, H. B., Sheng, Z. M., Lu, Z. S., Song, Q. L., & Li, C. M. (2010). Layered graphene/quantum dots for photovoltaic devices. *Angewandte Chemie International Edition*, *49*(17), 3014–3017.

226. Guo, S. R., Lin, J., Penchev, M., Yengel, E., Ghazinejad, M., Ozkan, C. S., & Ozkan, M. (2011). Label free DNA detection using large area graphene based field effect transistor biosensors. *Journal of Nanoscience and Nanotechnology*, *11*(6), 5258–5263.

227. Guo, S., Ghazinejad, M., Qin, X., Sun, H., Wang, W., Zaera, F., & Ozkan, C. S. (2012). Tuning Electron Transport in Graphene-Based Field-Effect Devices using Block Co-polymers. *Small*, *8*(7), 1073–1080.

228. Guo, X., Wang, C. F., Yu, Z. Y., Chen, L., & Chen, S. (2012). Facile access to versatile fluorescent carbon dots toward light-emitting diodes. *Chemical Communications*, *48*(21), 2692–2694.

229. Gur, I., Fromer, N. A., Chen, C. P., Kanaras, A. G., & Alivisatos, A. P. (2007). Hybrid solar cells with prescribed nanoscale morphologies based on hyperbranched semiconductor nanocrystals. *Nano letters*, *7*(2), 409–414.

230. Han, M. Y., Özyilmaz, B., Zhang, Y., & Kim, P. (2007). Energy band-gap engineering of graphene nanoribbons. *Physical Review Letters*, *98*(20), 206805.

231. Hanna, M. C., & Nozik, A. J. (2006). Solar conversion efficiency of photovoltaic and photoelectrolysis cells with carrier multiplication absorbers. *Journal of Applied Physics*, *100*(7), 074510.

232. Hanna, M. C., Beard, M. C., & Nozik, A. J. (2012). Effect of solar concentration on the thermodynamic power conversion efficiency of quantum-dot solar cells exhibiting multiple exciton generation. *The Journal of Physical Chemistry Letters*, *3*(19), 2857–2862.

233. Hao, X. J., Cho, E. C., Scardera, G., Shen, Y. S., Bellet-Amalric, E., Bellet, D., & Green, M. A. (2009). Phosphorus-doped silicon quantum dots for all-silicon quantum dot tandem solar cells. *Solar Energy Materials and Solar Cells*, *93*(9), 1524–1530.

234. Hegedus, S. S., & Shafarman, W. N. (2004). Thin-film solar cells: device measurements and analysis. *Progress in Photovoltaics: Research and Applications*, *12*(2-3), 155–176.

235. Hegedus, S., Desai, D., & Thompson, C. (2007). Voltage dependent photocurrent collection in CdTe/CdS solar cells. *Progress in Photovoltaics: Research and Applications*, *15*(7), 587–602.

236. Hetsch, F., Xu, X., Wang, H., Kershaw, S. V., & Rogach, A. L. (2011). Semiconductor nanocrystal quantum dots as solar cell components and photosensitizers: material, charge transfer, and separation aspects of some device topologies. *The Journal of Physical Chemistry Letters*, *2*(15), 1879–1887.

237. Hezam, M., Tabet, N., & Mekki, A. (2010). Synthesis and characterization of DC magnetron sputtered ZnO thin films under high working pressures. *Thin Solid Films*, *518*(24), e161–e164.

238. Hill, M. T., Marell, M., Leong, E. S., Smalbrugge, B., Zhu, Y., Sun, M., & Nötzel, R. (2009). Lasing in metal-insulator-metal sub-wavelength plasmonic waveguides. *Optics Express*, *17*(13), 11107–11112.

239. Hindson, J. C., Saghi, Z., Hernandez-Garrido, J. C., Midgley, P. A., & Greenham, N. C. (2011). Morphological study of nanoparticle−polymer solar cells using high-angle annular dark-field electron tomography. *Nano Letters*, *11*(2), 904–909.

240. Hines, M. A., & Guyot-Sionnest, P. (1996). Synthesis and characterization of strongly luminescing ZnS-capped CdSe nanocrystals. *The Journal of Physical Chemistry*, *100*(2), 468–471.

241. Hirschman, K. D., Tsybeskov, L., Duttagupta, S. P., & Fauchet, P. M. (1997, May). Integration of silicon electroluminescent devices with

silicon microelectronics. In *Conference on Lasers and Electro-Optics* (p. CThX7). Optical Society of America.

242. Hodes, G. (2008). Comparison of dye-and semiconductor-sensitized porous nanocrystalline liquid junction solar cells. *The Journal of Physical Chemistry C, 112*(46), 17778–17787.

243. Hohng, S., & Ha, T. (2004). Near-complete suppression of quantum dot blinking in ambient conditions. *Journal of the American Chemical Society, 126*(5), 1324–1325.

244. Hong, S. H., Park, J. H., Shin, D. H., Kim, C. O., Choi, S. H., & Kim, K. J. (2010). Doping-and size-dependent photovoltaic properties of p-type Si-quantum-dot heterojunction solar cells: correlation with photoluminescence. *Applied Physics Letters, 97*(7), 072108.

245. Honold, A., Schultheis, L., Kuhl, J., & Tu, C. W. (1989). Collision broadening of two-dimensional excitons in a GaAs single quantum well. *Physical Review B, 40*(9), 6442.

246. Hopfer, F., Kaiander, I., Lochmann, A., Mutig, A., Bognar, S., Kuntz, M., & Bimberg, D. (2006). Vertical-cavity surface-emitting quantum-dot laser with low threshold current grown by metal-organic vapor phase epitaxy. *Applied Physics Letters, 89*(6), 061105.

247. Hopfer, F., Mutig, A., Fiol, G., Kuntz, M., Shchukin, V. A., Haisler, V. A., & Livshits, D. A. (2007). 20 Gb/s 85$^{\circ} $ C Error-Free Operation of VCSELs Based on Submonolayer Deposition of Quantum Dots. *IEEE Journal of Selected Topics in Quantum Electronics, 13*(5), 1302–1308.

248. Hu, L., Zhao, Y. L., Ryu, K., Zhou, C., Stoddart, J. F., & Grüner, G. (2008). Light-Induced Charge Transfer in Pyrene/CdSe-SWNT Hybrids. *Advanced Materials, 20*(5), 939–946.

249. Hu, S. L., Niu, K. Y., Sun, J., Yang, J., Zhao, N. Q., & Du, X. W. (2009). One-step synthesis of fluorescent carbon nanoparticles by laser irradiation. *Journal of Materials Chemistry, 19*(4), 484–488.

250. Hu, Y., Diao, X., Wang, C., Hao, W., & Wang, T. (2004). Effects of heat treatment on properties of ITO films prepared by rf magnetron sputtering. *Vacuum, 75*(2), 183–188.

251. Hubbard, S. M., Cress, C. D., Bailey, C. G., Raffaelle, R. P., Bailey, S. G., & Wilt, D. M. (2008). Effect of strain compensation on quantum dot enhanced GaAs solar cells. *Applied Physics Letters, 92*(12), 123512.

252. Hudgings, J. A., Stone, R. J., Lim, S. F., Li, G. S., Yuen, W., Lau, K.

Y., & Chang-Hasnain, C. J. (1998). The physics of negative differential resistance of an intracavity voltage-controlled absorber in a vertical-cavity surface-emitting laser. *Applied Physics Letters*, *73*(13), 1796–1798.

253. Hughes, B. K., Luther, J. M., & Beard, M. C. (2012). The subtle chemistry of colloidal, quantum-confined semiconductor nanostructures. *ACS Nano*, *6*(6), 4573–4579.

254. Hughes, B. K., Ruddy, D. A., Blackburn, J. L., Smith, D. K., Bergren, M. R., Nozik, A. J., & Beard, M. C. (2012). Control of PbSe quantum dot surface chemistry and photophysics using an alkylselenide ligand. *ACS Nano*, *6*(6), 5498–5506.

255. Hungría, A. B., Juárez, B. H., Klinke, C., Weller, H., & Midgley, P. A. (2008). 3-D characterization of CdSe nanoparticles attached to carbon nanotubes. *Nano Research*, *1*(1), 89–97.

256. Huyet, G., Hegarty, S. P., O'Brien, D., Uskov, A. V., Melnik, S., Carroll, O., & Ustinov, V. M. (2004, September). Damping and feedback characteristics of quantum dot semiconductor lasers. In *Semiconductor Lasers and Laser Dynamics* (Vol. 5452, pp. 509–518). International Society for Optics and Photonics.

257. Huyet, G., O'Brien, D., Hegarty, S. P., McInerney, J. G., Uskov, A. V., Bimberg, D., & Kovsh, A. R. (2004). Quantum dot semiconductor lasers with optical feedback. *Physica Status Solidi (a)*, *201*(2), 345–352.

258. Huynh, W. U., Dittmer, J. J., & Alivisatos, A. P. (2002). Hybrid nanorod-polymer solar cells. *Science*, *295*(5564), 2425–2427.

259. Hyun, B. R., Zhong, Y. W., Bartnik, A. C., Sun, L., Abruna, H. D., Wise, F. W., & Borrelli, N. F. (2008). Electron injection from colloidal PbS quantum dots into titanium dioxide nanoparticles. *ACS Nano*, *2*(11), 2206–2212.

260. Iacono, F., Palencia, C., de la Cueva, L., Meyns, M., Terracciano, L., Vollmer, A., & Otero, R. (2013). Interfacing quantum dots and graphitic surfaces with chlorine atomic ligands. *ACS Nano*, *7*(3), 2559–2565.

261. Im, J. H., Lee, C. R., Lee, J. W., Park, S. W., & Park, N. G. (2011). 6.5% efficient perovskite quantum-dot-sensitized solar cell. *Nanoscale*, *3*(10), 4088–4093.

262. Inamdar, S. N., Ingole, P. P., & Haram, S. K. (2008). Determination of Band Structure Parameters and the Quasi-Particle Gap of CdSe

Quantum Dots by Cyclic Voltammetry. *ChemPhysChem*, *9*(17), 2574–2579.

263. Ip, A. H., Thon, S. M., Hoogland, S., Voznyy, O., Zhitomirsky, D., Debnath, R., & Kemp, K. W. (2012). Hybrid passivated colloidal quantum dot solids. *Nature nanotechnology*, *7*(9), 577.
264. Isborn, C. M., Kilina, S. V., Li, X., & Prezhdo, O. V. (2008). Generation of multiple excitons in PbSe and CdSe quantum dots by direct photoexcitation: first-principles calculations on small PbSe and CdSe clusters. *The Journal of Physical Chemistry C*, *112*(47), 18291–18294.
265. Jaeger, H. M., Fischer, S., & Prezhdo, O. V. (2012). Decoherence-induced surface hopping. *The Journal of Chemical Physics*, *137*(22), 22A545.
266. Jarosz, M. V., Porter, V. J., Fisher, B. R., Kastner, M. A., & Bawendi, M. G. (2004). Photoconductivity studies of treated CdSe quantum dot films exhibiting increased exciton ionization efficiency. *Physical Review B*, *70*(19), 195327.
267. Jasieniak, J., Califano, M., & Watkins, S. E. (2011). Size-dependent valence and conduction band-edge energies of semiconductor nanocrystals. *ACS Nano*, *5*(7), 5888–5902.
268. Javaux, C., Mahler, B., Dubertret, B., Shabaev, A., Rodina, A. V., Efros, A. L., & Biadala, L. (2013). Thermal activation of non-radiative Auger recombination in charged colloidal nanocrystals. *Nature Nanotechnology*, *8*(3), 206.
269. Jean, J., Chang, S., Brown, P. R., Cheng, J. J., Rekemeyer, P. H., Bawendi, M. G., ... & Bulović, V. (2013). ZnO nanowire arrays for enhanced photocurrent in PbS quantum dot solar cells. *Advanced Materials*, *25*(20), 2790–2796.
270. Jeffries, A. M., Mamidanna, A., Ding, L., Hildreth, O., & Bertoni, M. I. (2016, June). Reactive silver ink as front contacts for high efficiency silicon heterojunction solar cells. In *Photovoltaic Specialists Conference (PVSC), 2016 IEEE 43rd* (pp. 1984–1988). IEEE.
271. Jestin, Y., Pucker, G., Ghulinyan, M., Ferrario, L., Bellutti, P., Picciotto, A., & Pavesi, L. (2010, August). Silicon solar cells with nano-crystalline silicon down shifter: experiment and modeling. In *Next Generation (Nano) Photonic and Cell Technologies for Solar Energy Conversion* (Vol. 7772, p. 77720B). International Society for Optics and Photonics.
272. Ji, X., Makarov, N. S., Wang, W., Palui, G., Robel, I., & Mattoussi,

H. (2015). Tuning the Redox Coupling between Quantum Dots and Dopamine in Hybrid Nanoscale Assemblies. *The Journal of Physical Chemistry C*, *119*(6), 3388–3399.

273. Jia, X., Li, J., & Wang, E. (2012). One-pot green synthesis of optically pH-sensitive carbon dots with upconversion luminescence. *Nanoscale*, *4*(18), 5572–5575.

274. Jiao, S., Shen, Q., Mora-Seró, I., Wang, J., Pan, Z., Zhao, K., & Bisquert, J. (2015). Band engineering in core/shell ZnTe/CdSe for photovoltage and efficiency enhancement in exciplex quantum dot sensitized solar cells. *ACS Nano*, *9*(1), 908–915.

275. Joergensen, C., Danielsen, S. L., Stubkjaer, K. E., Schilling, M., Daub, K., Doussiere, P., & Vaa, M. (1997). All-optical wavelength conversion at bit rates above 10 Gb/s using semiconductor optical amplifiers. *IEEE Journal of Selected Topics in Quantum Electronics*, *3*(5), 1168–1180.

276. Johnston, K. (2008). Solution-processed Schottky-quantum Dot photovoltaics for efficient infrared power conversion (Doctoral dissertation).

277. Johnston, K. W., Pattantyus-Abraham, A. G., Clifford, J. P., Myrskog, S. H., MacNeil, D. D., Levina, L., & Sargent, E. H. (2008). Schottky-quantum dot photovoltaics for efficient infrared power conversion. *Applied Physics Letters*, *92*(15), 151115.

278. Johnston, K. W., Pattantyus-Abraham, A. G., Clifford, J. P., Myrskog, S. H., Hoogland, S., Shukla, H., & Sargent, E. H. (2008). Efficient Schottky-quantum-dot photovoltaics: The roles of depletion, drift, and diffusion. *Applied Physics Letters*, *92*(12), 122111.

279. Juarez-Perez, E. J., Wuβler, M., Fabregat-Santiago, F., Lakus-Wollny, K., Mankel, E., Mayer, T., & Mora-Sero, I. (2014). Role of the selective contacts in the performance of lead halide perovskite solar cells. *The Journal of Physical Chemistry Letters*, *5*(4), 680–685.

280. Juárez, B. H., Klinke, C., Kornowski, A., & Weller, H. (2007). Quantum dot attachment and morphology control by carbon nanotubes. *Nano Letters*, *7*(12), 3564–3568.

281. Juarez, B. H., Meyns, M., Chanaewa, A., Cai, Y., Klinke, C., & Weller, H. (2008). Carbon supported CdSe nanocrystals. *Journal of the American Chemical Society*, *130*(46), 15282–15284.

282. Jun, H. K. (2014). Parametric optimization in quantum dot-sensitized solar cells for efficiency enhancement/Jun Hieng Kiat (Doctoral

dissertation, University of Malaya).

283. Jun, H. K., Careem, M. A., & Arof, A. K. (2013). Quantum dot-sensitized solar cells—perspective and recent developments: a review of Cd chalcogenide quantum dots as sensitizers. *Renewable and Sustainable Energy Reviews*, *22*, 148–167.

284. Kaiser, E. T., & Kezdy, F. J. (1984). Amphiphilic secondary structure: design of peptide hormones. *Science*, *223*(4633), 249–255.

285. Kalyuzhny, G., & Murray, R. W. (2005). Ligand effects on optical properties of CdSe nanocrystals. *The Journal of Physical Chemistry B*, *109*(15), 7012–7021.

286. Kamat, P. V. (2007). Meeting the clean energy demand: nanostructure architectures for solar energy conversion. *The Journal of Physical Chemistry C*, *111*(7), 2834–2860.

287. Kamat, P. V. (2008). CENTENNIAL FEATURE ARTICLE. *J. Phys. Chem*, *100*(112), 18737–18753.

288. Kamat, P. V. (2008). Quantum dot solar cells. Semiconductor nanocrystals as light harvesters. *The Journal of Physical Chemistry C*, *112*(48), 18737–18753.

289. Kamath, K., Bhattacharya, P., Sosnowski, T., Norris, T., & Phillips, J. (1996). Room-temperature operation of In0.4Ga0.6As/GaAs self-organised quantum dot lasers. *Electronics Letters*, *32*(15), 1374–1375.

290. Kanemitsu, Y. (1994). Luminescence properties of nanometer-sized Si crystallites: Core and surface states. *Physical Review B*, *49*(23), 16845.

291. Kanemitsu, Y. (2013). Multiple exciton generation and recombination in carbon nanotubes and nanocrystals. *Accounts of Chemical Research*, *46*(6), 1358–1366.

292. Kanemitsu, Y., Okamoto, S., Otobe, M., & Oda, S. (1997). Photoluminescence mechanism in surface-oxidized silicon nanocrystals. *Physical Review B*, *55*(12), R7375.

293. Kelzenberg, M. D., Boettcher, S. W., Petykiewicz, J. A., Turner-Evans, D. B., Putnam, M. C., Warren, E. L., & Atwater, H. A. (2010). Enhanced absorption and carrier collection in Si wire arrays for photovoltaic applications. *Nature Materials*, *9*(3), 239.

294. Khan, S. A., & Bowden, A. K. E. (2016). Colloidal Quantum Dots for Cost-effective, Miniaturized, and Simple Spectrometers. *Clinical Chemistry*, *62*(4), 548–550.

295. Khoshman, J. M., & Kordesch, M. E. (2006). Optical properties of

a-HfO2 thin films. *Surface and Coatings Technology*, *201*(6), 3530–3535.

296. Khriachtchev, L., Kilpelä, O., Karirinne, S., Keränen, J., & Lepistö, T. (2001). Substrate-dependent crystallization and enhancement of visible photoluminescence in thermal annealing of Si/SiO 2 superlattices. *Applied Physics Letters*, *78*(3), 323–325.
297. Kim, J., Laemmlin, M., Meuer, C., Bimberg, D., & Eisenstein, G. (2009). Theoretical and experimental study of high-speed small-signal cross-gain modulation of quantum-dot semiconductor optical amplifiers. *IEEE Journal of Quantum Electronics*, *45*(3), 240–248.
298. Kim, J., Meuer, C., Bimberg, D., & Eisenstein, G. (2009). Role of carrier reservoirs on the slow phase recovery of quantum dot semiconductor optical amplifiers. *Applied Physics Letters*, *94*(4), 041112.
299. Kim, L., Anikeeva, P. O., Coe-Sullivan, S. A., Steckel, J. S., Bawendi, M. G., & Bulovic, V. (2008). Contact printing of quantum dot light-emitting devices. *Nano Letters*, *8*(12), 4513–4517.
300. Kim, M. R., & Ma, D. (2014). Quantum-dot-based solar cells: recent advances, strategies, and challenges. *The Journal of Physical Chemistry Letters*, *6*(1), 85–99.
301. Kim, S., Lim, Y. T., Soltesz, E. G., De Grand, A. M., Lee, J., Nakayama, A., & Cohn, L. H. (2004). Near-infrared fluorescent type II quantum dots for sentinel lymph node mapping. *Nature Biotechnology*, *22*(1), 93.
302. Kim, S., Marshall, A. R., Kroupa, D. M., Miller, E. M., Luther, J. M., Jeong, S., & Beard, M. C. (2015). Air-stable and efficient PbSe quantum-dot solar cells based upon ZnSe to PbSe cation-exchanged quantum dots. *ACS Nano*, *9*(8), 8157–8164.
303. Kim, Y. T., Han, J. H., Hong, B. H., & Kwon, Y. U. (2010). Electrochemical Synthesis of CdSe Quantum-Dot Arrays on a Graphene Basal Plane Using Mesoporous Silica Thin-Film Templates. *Advanced Materials*, *22*(4), 515–518.
304. Kinder, E., Moroz, P., Diederich, G., Johnson, A., Kirsanova, M., Nemchinov, A., & Zamkov, M. (2011). Fabrication of all-inorganic nanocrystal solids through matrix encapsulation of nanocrystal arrays. *Journal of the American Chemical Society*, *133*(50), 20488–20499.
305. Kirstaedter, N., Ledentsov, N. N., Grundmann, M., Bimberg, D., Ustinov, V. M., Ruvimov, S. S., & Werner, P. (1994). Low threshold,

large T/sub o/injection laser emission from (InGa) As quantum dots. *Electronics Letters*, *30*(17), 1416–1417.

306. Kirstaedter, N., Schmidt, O. G., Ledentsov, N. N., Bimberg, D., Ustinov, V. M., Egorov, A. Y., & Alferov, Z. I. (1996). Gain and differential gain of single layer InAs/GaAs quantum dot injection lasers. *Applied Physics Letters*, *69*(9), 1226–1228.
307. Klampaftis, E., Ross, D., McIntosh, K. R., & Richards, B. S. (2009). Enhancing the performance of solar cells via luminescent down-shifting of the incident spectrum: A review. *Solar Energy Materials and Solar Cells*, *93*(8), 1182–1194.
308. Klauk, H. (Ed.). (2012). *Organic electronics II: More Materials and Applications (Vol. 2)*. John Wiley & Sons.
309. Klem, E. J., MacNeil, D. D., Cyr, P. W., Levina, L., & Sargent, E. H. (2007). Efficient solution-processed infrared photovoltaic cells: Planarized all-inorganic bulk heterojunction devices via inter-quantum-dot bridging during growth from solution. *Applied physics Letters*, *90*(18), 183113.
310. Klimov, V. I. (2000). Optical nonlinearities and ultrafast carrier dynamics in semiconductor nanocrystals.
311. Klimov, V. I., Baker, T. A., Lim, J., Velizhanin, K. A., & McDaniel, H. (2016). Quality factor of luminescent solar concentrators and practical concentration limits attainable with semiconductor quantum dots. *ACS Photonics*, *3*(6), 1138–1148.
312. Klimov, V. I., Mikhailovsky, A. A., McBranch, D. W., Leatherdale, C. A., & Bawendi, M. G. (2000). Quantization of multiparticle Auger rates in semiconductor quantum dots. *Science*, *287*(5455), 1011–1013.
313. Ko, D. K., Brown, P. R., Bawendi, M. G., & Bulović, V. (2014). p-i-n Heterojunction Solar Cells with a Colloidal Quantum-Dot Absorber Layer. *Advanced Materials*, *26*(28), 4845–4850.
314. Kobayashi, S., Tani, Y., & Kawazoe, H. (2007). Quantum dot activated all-inorganic electroluminescent device fabricated using solution-synthesized CdSe/ZnS nanocrystals. *Japanese Journal of Applied Physics*, *46*(10L), L966.
315. Koleilat, G. I., Levina, L., Shukla, H., Myrskog, S. H., Hinds, S., Pattantyus-Abraham, A. G., & Sargent, E. H. (2008). Efficient, stable infrared photovoltaics based on solution-cast colloidal quantum dots. *ACS Nano*, *2*(5), 833–840.

316. Kolny-Olesiak, J., & Weller, H. (2013). Synthesis and application of colloidal CuInS2 semiconductor nanocrystals. *ACS Applied Materials & Interfaces*, *5*(23), 12221–12237.

317. Kolodinski, S., Werner, J. H., Wittchen, T., & Queisser, H. J. (1993). Quantum efficiencies exceeding unity due to impact ionization in silicon solar cells. *Applied Physics Letters*, *63*(17), 2405–2407.

318. Kongkanand, A., Tvrdy, K., Takechi, K., Kuno, M., & Kamat, P. V. (2008). Quantum dot solar cells. Tuning photoresponse through size and shape control of CdSe– TiO2 architecture. *Journal of the American Chemical Society*, *130*(12), 4007–4015.

319. Konstantatos, G., Badioli, M., Gaudreau, L., Osmond, J., Bernechea, M., De Arquer, F. P. G., & Koppens, F. H. (2012). Hybrid graphene–quantum dot phototransistors with ultrahigh gain. *Nature Nanotechnology*, *7*(6), 363.

320. Konstantatos, G., Howard, I., Fischer, A., Hoogland, S., Clifford, J., Klem, E., & Sargent, E. H. (2006). Ultrasensitive solution-cast quantum dot photodetectors. *Nature*, *442*(7099), 180.

321. Kovalenko, M. V., Scheele, M., & Talapin, D. V. (2009). Colloidal nanocrystals with molecular metal chalcogenide surface ligands. *Science*, *324*(5933), 1417–1420.

322. Kovalev, D., Heckler, H., Ben-Chorin, M., Polisski, G., Schwartzkopff, M., & Koch, F. (1998). Breakdown of the k-conservation rule in Si nanocrystals. *Physical Review Letters*, *81*(13), 2803.

323. Kramer, I. J. (2012). *Architecting the Optics, Energetics and Geometry of Colloidal Quantum Dot Photovoltaics* (Doctoral dissertation).

324. Kramer, I. J., & Sargent, E. H. (2011). Colloidal quantum dot photovoltaics: a path forward. *ACS nano*, *5*(11), 8506–8514.

325. Kramer, I. J., & Sargent, E. H. (2013). The architecture of colloidal quantum dot solar cells: materials to devices. *Chemical Reviews*, *114*(1), 863–882.

326. Kramer, I., Debnath, R., Pattantyus-Abraham, A. G., Barkhouse, A. R., Wang, X., Levina, L., & Lu, Z. H. (2010, October). Depleted Heterojunction Colloidal Quantum Dot Solar Cells Employing Low-Cost Metal Contacts. In *Frontiers in Optics* (p. FMA2). Optical Society of America.

327. Kraus, R. M., Lagoudakis, P. G., Rogach, A. L., Talapin, D. V., Weller, H., Lupton, J. M., & Feldmann, J. (2007). Room-temperature exciton

storage in elongated semiconductor nanocrystals. *Physical Review Letters*, *98*(1), 017401.

328. Kriegel, I., Jiang, C., Rodríguez-Fernández, J., Schaller, R. D., Talapin, D. V., Da Como, E., & Feldmann, J. (2012). Tuning the excitonic and plasmonic properties of copper chalcogenide nanocrystals. *Journal of the American Chemical Society*, *134*(3), 1583–1590.
329. Kucur, E., Riegler, J., Urban, G. A., & Nann, T. (2003). Determination of quantum confinement in CdSe nanocrystals by cyclic voltammetry. *The Journal of Chemical Physics*, *119*(4), 2333–23.
330. Kumar, S., & Scholes, G. D. (2008). Colloidal nanocrystal solar cells. *Microchemical Acta*, *160*(3), 315–325.
331. Kuno, M., Lee, J. K., Dabbousi, B. O., Mikulec, F. V., & Bawendi, M. G. (1997). The band edge luminescence of surface modified CdSe nanocrystallites: Probing the luminescing state. *The Journal of Chemical Physics*, *106*(23), 9869–9882.
332. Kuntz, M., Fiol, G., Laemmlin, M., Meuer, C., & Bimberg, D. (2007). High-speed mode-locked quantum-dot lasers and optical amplifiers. *Proceedings of the IEEE*, *95*(9), 1767–1778.
333. Kwak, J., Bae, W. K., Lee, D., Park, I., Lim, J., Park, M., & Lee, S. (2012). Bright and efficient full-color colloidal quantum dot light-emitting diodes using an inverted device structure. *Nano Letters*, *12*(5), 2362–2366.
334. Kwon, W., Do, S., Lee, J., Hwang, S., Kim, J. K., & Rhee, S. W. (2013). Freestanding luminescent films of nitrogen-rich carbon nanodots toward large-scale phosphor-based white-light-emitting devices. *Chemistry of Materials*, *25*(9), 1893–1899.
335. Kyle, J. R., Guvenc, A., Wang, W., Ghazinejad, M., Lin, J., Guo, S., & Ozkan, M. (2011). Centimeter-Scale High-Resolution Metrology of Entire CVD-Grown Graphene Sheets. *Small*, *7*(18), 2599–2606.
336. Lambert, K., Dendooven, J., Detavernier, C., & Hens, Z. (2010). Embedding quantum dot monolayers in Al_2O_3 using atomic layer deposition. *Chemistry of Materials*, *23*(2), 126–128.
337. Lan, X., Masala, S., & Sargent, E. H. (2014). Charge-extraction strategies for colloidal quantum dot photovoltaics. *Nature Materials*, *13*(3), 233.
338. Lan, X., Voznyy, O., García de Arquer, F. P., Liu, M., Xu, J., Proppe, A. H., & Yang, Z. (2016). 10.6% certified colloidal quantum dot solar

cells via solvent-polarity-engineered halide passivation. *Nano Letters*, *16*(7), 4630–4634.

339. Lan, X., Voznyy, O., Kiani, A., García de Arquer, F. P., Abbas, A. S., Kim, G. H., & Yuan, M. (2016). Passivation using molecular halides increases quantum dot solar cell performance. *Advanced Materials*, *28*(2), 299–304.
340. Law, M., Beard, M. C., Choi, S., Luther, J. M., Hanna, M. C., & Nozik, A. J. (2008). Determining the internal quantum efficiency of PbSe nanocrystal solar cells with the aid of an optical model. *Nano Letters*, *8*(11), 3904–3910.
341. Law, M., Luther, J. M., Beard, M. C., Choi, S., & Nozik, A. J. (2009, June). Solar cells based on colloidal quantum dot solids: seeking enhanced photocurrent. In *Photovoltaic Specialists Conference (PVSC), 2009 34th IEEE* (pp. 002068–002073). IEEE.
342. Le Donne, A., Acciarri, M., Narducci, D., Marchionna, S., & Binetti, S. (2009). Encapsulating Eu3+ complex doped layers to improve Si-based solar cell efficiency. *Progress in Photovoltaics: Research and Applications*, *17*(8), 519–525.
343. Le Donne, A., Dilda, M., Crippa, M., Acciarri, M., & Binetti, S. (2011). Rare earth organic complexes as down-shifters to improve Si-based solar cell efficiency. *Optical Materials*, *33*(7), 1012–1014.
344. Leatherdale, C. A., Woo, W. K., Mikulec, F. V., & Bawendi, M. G. (2002). On the absorption cross section of CdSe nanocrystal quantum dots. *The Journal of Physical Chemistry B*, *106*(31), 7619–7622.
345. Ledentsov, N. N., Grundmann, M., Heinrichsdorff, F., Bimberg, D., Ustinov, V. M., Zhukov, A. E., & Lott, J. A. (2000). Quantum-dot heterostructure lasers. *IEEE Journal of Selected Topics in Quantum Electronics*, *6*(3), 439–451.
346. Ledentsov, N. N., Hopfer, F., & Bimberg, D. (2007). High-speed quantum-dot vertical-cavity surface-emitting lasers. *Proceedings of the IEEE*, *95*(9), 1741–1756.
347. Ledentsov, N. N., Ledentsov, N., Agustin, M., Kropp, J. R., & Shchukin, V. A. (2017). Application of nanophotonics to the next generation of surface-emitting lasers. *Nanophotonics*, *6*(5), 813–829.
348. Ledentsov, N. N., Ustinov, V. M., Egorov, A. Y., Zhukov, A. E., Maksimov, M. V., Tabatadze, I. G., & Kop’ev, P. S. (1994). Optical properties of heterostructures with InGaAs-GaAs quantum clusters.

Semiconductors, *28*(8), 832–834.

349. Lee, D. C., Robel, I., Pietryga, J. M., & Klimov, V. I. (2010). Infrared-active heterostructured nanocrystals with ultralong carrier lifetimes. *Journal of the American Chemical Society*, *132*(29), 9960–9962.

350. Lee, D., Kim, W. D., Lee, S., Bae, W. K., Lee, S., & Lee, D. C. (2015). Direct Cd-to-Pb exchange of CdSe nanorods into PbSe/CdSe axial heterojunction nanorods. *Chemistry of Materials*, *27*(15), 5295–5304.

351. Lee, H. J., Chen, P., Moon, S. J., Sauvage, F., Sivula, K., Bessho, T., & Grätzel, M. (2009). Regenerative PbS and CdS quantum dot sensitized solar cells with a cobalt complex as hole mediator. *Langmuir*, 25(13), 7602–7608.

352. Lee, H. J., Yum, J. H., Leventis, H. C., Zakeeruddin, S. M., Haque, S. A., Chen, P., & Nazeeruddin, M. K. (2008). CdSe quantum dot-sensitized solar cells exceeding efficiency 1% at full-sun intensity. *The Journal of Physical Chemistry C*, *112*(30), 11600–11608.

353. Lee, H., Leventis, H. C., Moon, S. J., Chen, P., Ito, S., Haque, S. A., & Grätzel, M. (2009). PbS and CdS quantum dot-sensitized solid-state solar cells: "Old concepts, new results." *Advanced Functional Materials*, *19*(17), 2735–2742.

354. Lee, H., Wang, M., Chen, P., Gamelin, D. R., Zakeeruddin, S. M., Gratzel, M., & Nazeeruddin, M. K. (2009). Efficient CdSe quantum dot-sensitized solar cells prepared by an improved successive ionic layer adsorption and reaction process. *Nano Letters*, *9*(12), 4221–4227.

355. Lee, I. J., Paik, U., & Park, J. G. (2013). Effects of metallic contaminant type and concentration on photovoltaic performance degradation of p-type silicon solar cells. *Journal of the Korean Physical Society*, *63*(1), 47–52.

356. Lee, M. M., Teuscher, J., Miyasaka, T., Murakami, T. N., & Snaith, H. J. (2012). Efficient hybrid solar cells based on meso-superstructured organometal halide perovskites. *Science*, 1228604.

357. Lee, W., Kang, S. H., Min, S. K., Sung, Y. E., & Han, S. H. (2008). Co-sensitization of vertically aligned TiO2 nanotubes with 2 different sizes of CdSe quantum dots for broad spectrum. *Electrochemistry Communications*, *10*(10), 1579–1582.

358. Lee, W., Lee, J., Lee, S., Yi, W., Han, S. H., & Cho, B. W. (2008). Enhanced charge collection and reduced recombination of Cd S/Ti O 2 quantum-dots sensitized solar cells in the presence of single-walled

carbon nanotubes. *Applied Physics Letters*, *92*(15), 153510.

359. Lee, W., Lee, J., Min, S. K., Park, T., Yi, W., & Han, S. H. (2009). Effect of single-walled carbon nanotube in PbS/TiO2 quantum dots-sensitized solar cells. *Materials Science and Engineering: B*, *156*(1–3), 48–51.
360. Lee, Y. L., & Lo, Y. S. (2009). Highly efficient quantum-dot-sensitized solar cell based on co-sensitization of CdS/CdSe. *Advanced Functional Materials*, *19*(4), 604–609.
361. Lee, Y. L., Chi, C. F., & Liau, S. Y. (2009). CdS/CdSe co-sensitized TiO2 photoelectrode for efficient hydrogen generation in a photoelectrochemical cell. *Chemistry of Materials*, *22*(3), 922–927.
362. Lee, Y. L., Huang, B. M., & Chien, H. T. (2008). Highly efficient CdSe-sensitized TiO2 photoelectrode for quantum-dot-sensitized solar cell applications. *Chemistry of Materials*, *20*(22), 6903–6905.
363. Lehmann, V., & Gösele, U. (1991). Porous silicon formation: A quantum wire effect. *Applied Physics Letters*, *58*(8), 856–858.
364. Lenz, A., Eisele, H., Becker, J., Ivanova, L., Lenz, E., Luckert, F., & Dähne, M. (2010). Atomic structure of buried InAs sub-monolayer depositions in GaAs. *Applied Physics Express*, *3*(10), 105602.
365. Leschkies, K. S., Beatty, T. J., Kang, M. S., Norris, D. J., & Aydil, E. S. (2009). Solar cells based on junctions between colloidal PbSe nanocrystals and thin ZnO films. *ACS nano*, *3*(11), 3638–3648.
366. Leschkies, K. S., Divakar, R., Basu, J., Enache-Pommer, E., Boercker, J. E., Carter, C. B., & Aydil, E. S. (2007). Photosensitization of ZnO nanowires with CdSe quantum dots for photovoltaic devices. *Nano Letters*, *7*(6), 1793–1798.
367. Leschkies, K. S., Kang, M. S., Aydil, E. S., & Norris, D. J. (2010). Influence of atmospheric gases on the electrical properties of PbSe quantum-dot films. *The Journal of Physical Chemistry C*, *114*(21), 9988–9996.
368. Li, B., Wang, L., Kang, B., Wang, P., & Qiu, Y. (2006). Review of recent progress in solid-state dye-sensitized solar cells. *Solar Energy Materials and Solar Cells*, *90*(5), 549–573.
369. Li, H., He, X., Kang, Z., Huang, H., Liu, Y., Liu, J., & Lee, S. T. (2010). Water-soluble fluorescent carbon quantum dots and photocatalyst design. *Angewandte Chemie International Edition*, *49*(26), 4430–4434.
370. Li, H., Wu, K., Lim, J., Song, H. J., & Klimov, V. I. (2016). Doctor-

blade deposition of quantum dots onto standard window glass for low-loss large-area luminescent solar concentrators. *Nature Energy*, *1*(12), 16157.

371. Li, H., Zhitomirsky, D., & Grossman, J. C. (2016). Tunable and energetically robust PbS nanoplatelets for optoelectronic applications. *Chemistry of Materials*, *28*(6), 1888–1896.

372. Li, Q., Zhao, J., Sun, B., Lin, B., Qiu, L., Zhang, Y., & Yan, F. (2012). High-Temperature Solid-State Dye-Sensitized Solar Cells Based on Organic Ionic Plastic Crystal Electrolytes. *Advanced Materials*, *24*(7), 945–950.

373. Li, X., Cai, W., An, J., Kim, S., Nah, J., Yang, D., & Tutuc, E. (2009). Banerjee., SK; Colombo, L.; Ruoff, RS. *Science*, *324*, 1312.

374. Li, X., Jia, Y., & Cao, A. (2009). Tailored single-walled carbon nanotube− CdS nanoparticle hybrids for tunable optoelectronic devices. *Acs Nano*, *4*(1), 506–512.

375. Li, X., Jia, Y., Wei, J., Zhu, H., Wang, K., Wu, D., & Cao, A. (2010). Solar cells and light sensors based on nanoparticle-grafted carbon nanotube films. *ACS Nano*, *4*(4), 2142–2148.

376. Li, X., Liu, Y., Song, X., Wang, H., Gu, H., & Zeng, H. (2015). Intercrossed Carbon Nanorings with Pure Surface States as Low-Cost and Environment-Friendly Phosphors for White-Light-Emitting Diodes. *Angewandte Chemie International Edition*, *54*(6), 1759–1764.

377. Li, X., Wang, X., Zhang, L., Lee, S., & Dai, H. (2008). Chemically derived, ultrasmooth graphene nanoribbon semiconductors. *Science*, *319*(5867), 1229–1232.

378. Li, X., Zhang, S., Kulinich, S. A., Liu, Y., & Zeng, H. (2014). Engineering surface states of carbon dots to achieve controllable luminescence for solid-luminescent composites and sensitive Be 2+ detection. *Scientific reports*, *4*, 4976.

379. Li, Y., Zhao, Y., Cheng, H., Hu, Y., Shi, G., Dai, L., & Qu, L. (2011). Nitrogen-doped graphene quantum dots with oxygen-rich functional groups. *Journal of the American Chemical Society*, *134*(1), 15–18.

380. Li, Z., Ji, Y., Xie, R., Grisham, S. Y., & Peng, X. (2011). Correlation of CdS nanocrystal formation with elemental sulfur activation and its implication in synthetic development. *Journal of the American Chemical Society*, *133*(43), 17248–17256.

381. Li, Z., Wu, P., Wang, C., Fan, X., Zhang, W., Zhai, X., & Hou, J. G.

(2011). Chemical Vapor Deposition Growth of Graphene using Other Hydrocarbon Sources. *arXiv Preprint arXiv:1101.5664.*

382. Lim, S. Y., Shen, W., & Gao, Z. (2015). Carbon quantum dots and their applications. *Chemical Society Reviews*, *44*(1), 362–381.
383. Lin, J., Zhong, J., Kyle, J. R., Penchev, M., Ozkan, M., & Ozkan, C. S. (2011). Molecular absorption and photodesorption in pristine and functionalized large-area graphene layers. *Nanotechnology*, *22*(35), 355701.
384. Lin, Z., Franceschetti, A., & Lusk, M. T. (2011). Size dependence of the multiple exciton generation rate in CdSe quantum dots. *ACS Nano*, *5*(4), 2503–2511.
385. Linnros, J., Lalic, N., Galeckas, A., & Grivickas, V. (1999). Analysis of the stretched exponential photoluminescence decay from nanometer-sized silicon crystals in SiO 2. *Journal of Applied Physics*, *86*(11), 6128–6134.
386. Liu, B., & Zeng, H. C. (2008). Carbon nanotubes supported mesoporous mesocrystals of anatase TiO2.*Chemistry of Materials*, *20*(8), 2711–2718.
387. Liu, C. Y. (2009). Hybrid Solar Cells from Polymers and Silicon Nanocrystals.
388. Liu, C. Y., & Kortshagen, U. R. (2010). A silicon nanocrystal Schottky junction solar cell produced from colloidal silicon nanocrystals. *Nanoscale Research Letters*, *5*(8), 1253.
389. Liu, C. Y., Holman, Z. C., & Kortshagen, U. R. (2008). Hybrid solar cells from P3HT and silicon nanocrystals. *Nano Letters*, *9*(1), 449–452.
390. Liu, L., Barber, G. D., Shuba, M. V., Yuwen, Y., Lakhtakia, A., Mallouk, T. E., & Mayer, T. S. (2016). Planar light concentration in micro-Si solar cells enabled by a metallic grating–photonic crystal architecture. *ACS Photonics*, *3*(4), 604–610.
391. Liu, L., Hensel, J., Fitzmorris, R. C., Li, Y., & Zhang, J. Z. (2009). Preparation and photoelectrochemical properties of CdSe/TiO_2 hybrid mesoporous structures. *The Journal of Physical Chemistry Letters*, *1*(1), 155–160.
392. Liu, M., de Arquer, F., Li, Y., Lan, X., Kim, G. H., Voznyy, O., & Kim, J. Y. (2016). Double-sided junctions enable high-performance colloidal-quantum-dot photovoltaics. *Advanced Materials*, *28*(21), 4142–4148.

393. Liu, M., Johnston, M. B., & Snaith, H. J. (2013). Efficient planar heterojunction perovskite solar cells by vapour deposition. *Nature, 501*(7467), 395.

394. Liu, M., Voznyy, O., Sabatini, R., de Arquer, F. P. G., Munir, R., Balawi, A. H., & Hoogland, S. (2017). Hybrid organic–inorganic inks flatten the energy landscape in colloidal quantum dot solids. *Nature Materials, 16*(2), 258.

395. Liu, R., Wu, D., Liu, S., Koynov, K., Knoll, W., & Li, Q. (2009). An aqueous route to multicolor photoluminescent carbon dots using silica spheres as carriers. *Angewandte Chemie, 121*(25), 4668–4671.

396. Liu, Y., Gibbs, M., Perkins, C. L., Tolentino, J., Zarghami, M. H., Bustamante Jr, J., & Law, M. (2011). Robust, functional nanocrystal solids by infilling with atomic layer deposition. *Nano Letters, 11*(12), 5349–5355.

397. Liu, Y., Gibbs, M., Puthussery, J., Gaik, S., Ihly, R., Hillhouse, H. W., & Law, M. (2010). Dependence of carrier mobility on nanocrystal size and ligand length in PbSe nanocrystal solids. *Nano letters, 10*(5), 1960–1969.

398. Liu, Z., Sun, Y., Yuan, J., Wei, H., Huang, X., Han, L., & Ma, W. (2013). High-Efficiency Hybrid Solar Cells Based on Polymer/PbSxSe1-x Nanocrystals Benefiting from Vertical Phase Segregation. *Advanced Materials, 25*(40), 5772–5778.

399. Lockwood, D. J., Lu, Z. H., & Baribeau, J. M. (1996). Quantum confined luminescence in Si/SiO 2 superlattices. *Physical Review Letters, 76*(3), 539.

400. Lott, J. A., Ledentsov, N. N., Ustinov, V. M., Egorov, A. Y., Zhukov, A. E., Kop'ev, P. S., ... & Bimberg, D. (1997). Vertical cavity lasers based on vertically coupled quantum dots. *Electronics Letters, 33*(13), 1150–1151.

401. Lu, C. Y., Chang, S. W., Chuang, S. L., Germann, T. D., & Bimberg, D. (2010). Metal-cavity surface-emitting microlaser at room temperature. *Applied Physics Letters, 96*(25), 251101.

402. Lu, C., Akey, A., Wang, W., & Herman, I. P. (2009). Versatile formation of CdSe nanoparticle-single walled carbon nanotube hybrid structures. *Journal of the American Chemical Society, 131*(10), 3446–3447.

403. Lu, G., Mao, S., Park, S., Ruoff, R. S., & Chen, J. (2009). Facile, noncovalent decoration of graphene oxide sheets with nanocrystals.

Nano Research, *2*(3), 192–200.

404. Lu, W., Qin, X., Liu, S., Chang, G., Zhang, Y., Luo, Y., & Sun, X. (2012). Economical, green synthesis of fluorescent carbon nanoparticles and their use as probes for sensitive and selective detection of mercury (II) ions. *Analytical Chemistry*, *84*(12), 5351–5357.
405. Luque, A., & Martí, A. (1997). Increasing the efficiency of ideal solar cells by photon induced transitions at intermediate levels. *Physical Review Letters*, *78*(26), 5014.
406. Luque, A., & Martí, A. (2010). Electron–phonon energy transfer in hot-carrier solar cells. *Solar Energy Materials and Solar Cells*, *94*(2), 287–296.
407. Luther, J. M., Gao, J., Lloyd, M. T., Semonin, O. E., Beard, M. C., & Nozik, A. J. (2010). Stability assessment on a 3% bilayer PbS/ZnO quantum dot heterojunction solar cell. *Advanced Materials*, *22*(33), 3704–3707.
408. Luther, J. M., Law, M., Beard, M. C., Song, Q., Reese, M. O., Ellingson, R. J., & Nozik, A. J. (2008). Schottky solar cells based on colloidal nanocrystal films. *Nano Letters*, *8*(10), 3488–3492.
409. Luther, J. M., Law, M., Song, Q., Perkins, C. L., Beard, M. C., & Nozik, A. J. (2008). Structural, optical, and electrical properties of self-assembled films of PbSe nanocrystals treated with 1, 2-ethanedithiol. *ACS Nano*, *2*(2), 271–280.
410. Ma, W., Luther, J. M., Zheng, H., Wu, Y., & Alivisatos, A. P. (2009). Photovoltaic Devices Employing Ternary Pbs X Se1-X Nanocrystals. *Nano Letters*, *9*(4), 1699–1703.
411. Maathuis, M. H., Colombo, D., Kalisch, M., & Bühlmann, P. (2010). Predicting causal effects in large-scale systems from observational data. *Nature Methods*, *7*(4), 247.
412. Maathuis, M. H., Kalisch, M., & Bühlmann, P. (2009). Estimating high-dimensional intervention effects from observational data. *The Annals of Statistics*, *37*(6A), 3133–3164.
413. Mahato, P., Monguzzi, A., Yanai, N., Yamada, T., & Kimizuka, N. (2015). Fast and long-range triplet exciton diffusion in metal–organic frameworks for photon upconversion at ultralow excitation power. *Nature Materials*, *14*(9), 924.
414. Mahler, B., Spinicelli, P., Buil, S., Quelin, X., Hermier, J. P., & Dubertret, B. (2008). Towards non-blinking colloidal quantum dots.

Nature Materials, *7*(8), 659.

415. Mangolini, L., Thimsen, E., & Kortshagen, U. (2005). High-yield plasma synthesis of luminescent silicon nanocrystals. *Nano Letters*, *5*(4), 655–659.

416. Mao, L. H., Tang, W. Q., Deng, Z. Y., Liu, S. S., Wang, C. F., & Chen, S. (2014). Facile access to white fluorescent carbon dots toward light-emitting devices. *Industrial & Engineering Chemistry Research*, *53*(15), 6417–6425.

417. Marconi, A., Anopchenko, A., Wang, M., Pucker, G., Bellutti, P., & Pavesi, L. (2009). High power efficiency in Si-nc/SiO 2 multilayer light emitting devices by bipolar direct tunneling. *Applied Physics Letters*, *94*(22), 221110.

418. Markovic, Z. M., Ristic, B. Z., Arsikin, K. M., Klisic, D. G., Harhaji-Trajkovic, L. M., Todorovic-Markovic, B. M., & Milivojevic, D. D. (2012). Graphene quantum dots as autophagy-inducing photodynamic agents. *Biomaterials*, *33*(29), 7084–7092.

419. Marshall, A. R., Beard, M. C., & Johnson, J. C. (2016). Nongeminate radiative recombination of free charges in cation-exchanged PbS quantum dot films. *Chemical Physics*, *471*, 75–80.

420. Marshall, A. R., Young, M. R., Nozik, A. J., Beard, M. C., & Luther, J. M. (2015). Exploration of metal chloride uptake for improved performance characteristics of PbSe quantum dot solar cells. *The Journal of Physical Chemistry Letters*, *6*(15), 2892–2899.

421. Martí, A., Antolín, E., Stanley, C. R., Farmer, C. D., López, N., Díaz, P., & Luque, A. (2006). Production of photocurrent due to intermediate-to-conduction-band transitions: a demonstration of a key operating principle of the intermediate-band solar cell. *Physical Review Letters*, *97*(24), 247701.

422. Maruyama, T., Enomoto, A., & Shirasawa, K. (2000). Solar cell module colored with fluorescent plate. *Solar Energy Materials and Solar Cells*, *64*(3), 269–278.

423. Mashford, B. S., Stevenson, M., Popovic, Z., Hamilton, C., Zhou, Z., Breen, C., & Kazlas, P. T. (2013). High-efficiency quantum-dot light-emitting devices with enhanced charge injection. *Nature Photonics*, *7*(5), 407.

424. May, K. J., Carlton, C. E., Stoerzinger, K. A., Risch, M., Suntivich, J., Lee, Y. L., & Shao-Horn, Y. (2012). Influence of oxygen evolution

during water oxidation on the surface of perovskite oxide catalysts. *The Journal of Physical Chemistry Letters*, *3*(22), 3264–3270.

425. Mayer, J. N., Simon, P., Philipps, N. S. H., Schlegl, T., & Senkpiel, C. (2015). Current and future cost of photovoltaics. *Long-term scenarios for market development, system prices and LCOE of utility-scale PV systems*.

426. McCaffrey, R. R., & Prasad, P. N. (1984). Organic-thin-film-coated solar cells: Energy transfer between surface pyrene molecules and the silicon semiconductor substrate. *Solar Cells*, *11*(4), 401–409.

427. McGuire, J. A., Joo, J., Pietryga, J. M., Schaller, R. D., & Klimov, V. I. (2008). New aspects of carrier multiplication in semiconductor nanocrystals. *Accounts of Chemical Research*, *41*(12), 1810–1819.

428. McIntosh, K. R., & Richards, B. S. (2006, May). Increased mc-Si module efficiency using fluorescent organic dyes: a ray-tracing study. In *Photovoltaic Energy Conversion, Conference Record of the 2006 IEEE 4th World Conference on* (*Vol. 2*, pp. 2108–2111). IEEE.

429. Meillaud, F., Shah, A., Droz, C., Vallat-Sauvain, E., & Miazza, C. (2006). Efficiency limits for single-junction and tandem solar cells. *Solar Energy Materials And Solar Cells*, *90*(18–19), 2952–2959.

430. Merrifield, R. E., Avakian, P., & Groff, R. P. (1969). Fission of singlet excitons into pairs of triplet excitons in tetracene crystals. *Chemical Physics Letters*, *3*(6), 386–388.

431. Meuer, C., Kim, J., Laemmlin, M., Liebich, S., Eisenstein, G., Bonk, R., & Bimberg, D. (2009). High-Speed Small-Signal Cross-Gain Modulation in Quantum-Dot Semiconductor Optical Amplifiers at 1.3$\mu $ m. *IEEE Journal of Selected Topics in Quantum Electronics*, *15*(3), 749–756.

432. Meuer, C., Schmidt-Langhorst, C., Bonk, R., Schmeckebier, H., Arsenijević, D., Fiol, G., & Bimberg, D. (2011). 80 Gb/s wavelength conversion using a quantum-dot semiconductor optical amplifier and optical filtering. *Optics Express*, *19*(6), 5134–5142.

433. Midgett, A. G., Hillhouse, H. W., Hughes, B. K., Nozik, A. J., & Beard, M. C. (2010). Flowing versus static conditions for measuring multiple exciton generation in PbSe quantum dots. *The Journal of Physical Chemistry C*, *114*(41), 17486–17500.

434. Miller, D. A. B., Chemla, D. S., Damen, T. C., Gossard, A. C., Wiegmann, W., Wood, T. H., & Burrus, C. A. (1985). Electric field

dependence of optical absorption near the band gap of quantum-well structures. *Physical Review B, 32*(2), 1043.

435. Miller, O. D., Yablonovitch, E., & Kurtz, S. R. (2012). Strong internal and external luminescence as solar cells approach the Shockley–Queisser limit. *IEEE Journal of Photovoltaics, 2*(3), 303–311.

436. Milliron, D. J. (2014). Quantum dot solar cells: The surface plays a core role. *Nature materials, 13*(8), 772.

437. Min, K. S., Shcheglov, K. V., Yang, C. M., Atwater, H. A., Brongersma, M. L., & Polman, A. (1996). Defect-related versus excitonic visible light emission from ion beam synthesized Si nanocrystals in SiO2. *Applied Physics Letters, 69*(14), 2033–2035.

438. Mirabella, S., Di Martino, G., Crupi, I., Gibilisco, S., Miritello, M., Lo Savio, R., & Priolo, F. (2010). Light absorption and electrical transport in Si: O alloys for photovoltaics. *Journal of Applied Physics, 108*(9), 093507.

439. Mirtchev, P., Henderson, E. J., Soheilnia, N., Yip, C. M., & Ozin, G. A. (2012). Solution phase synthesis of carbon quantum dots as sensitizers for nanocrystalline TiO 2 solar cells. *Journal of Materials Chemistry, 22*(4), 1265–1269.

440. Mishra, A., Fischer, M. K., & Bäuerle, P. (2009). Metal-free organic dyes for dye-sensitized solar cells: From structure: Property relationships to design rules. *Angewandte Chemie International Edition, 48*(14), 2474–2499.

441. Miyazaki, H. T., & Kurokawa, Y. (2006). Squeezing visible light waves into a 3-nm-thick and 55-nm-long plasmon cavity. *Physical Review Letters, 96*(9), 097401.

442. Mocatta, D., Cohen, G., Schattner, J., Millo, O., Rabani, E., & Banin, U. (2011). Heavily doped semiconductor nanocrystal quantum dots. *Science, 332*(6025), 77–81.

443. Mohan, S., Sreekala, C. O., Tom, A. E., Thomas, A., & Ison, V. V. (2016, March). Photovoltaic studies of PbSe quantum dot based solar cells. In *Electrical, Electronics, and Optimization Techniques (ICEEOT), International Conference on* (pp. 4625–4627). IEEE.

444. Moon, Y. K., Bang, B., Kim, S. H., Jeong, C. O., & Park, J. W. (2008). Effects of working pressure on the electrical and optical properties of aluminum-doped zinc oxide thin films. *Journal of Materials Science: Materials in Electronics, 19*(6), 528–532.

445. Mora-Seró, I., & Bisquert, J. (2010). Breakthroughs in the development of semiconductor-sensitized solar cells. *The Journal of Physical Chemistry Letters*, *1*(20), 3046–3052.

446. Moreels, I., Fritzinger, B., Martins, J. C., & Hens, Z. (2008). Surface chemistry of colloidal PbSe nanocrystals. *Journal of the American Chemical Society*, *130*(45), 15081–15086.

447. Morgenstern, F. S., Böhm, M. L., Kist, R. J., Sadhanala, A., Gélinas, S., Rao, A., & Greenham, N. C. (2016). Charge generation and electron-trapping dynamics in hybrid nanocrystal-polymer solar cells. *The Journal of Physical Chemistry C*, *120*(34), 19064–19069.

448. Moroz, P., Kholmicheva, N., Mellott, B., Liyanage, G., Rijal, U., Bastola, E., & Zamkov, M. (2013). Suppressed carrier scattering in CdS-encapsulated PbS nanocrystal films. *ACS Nano*, *7*(8), 6964–6977.

449. Moroz, P., Liyanage, G., Kholmicheva, N. N., Yakunin, S., Rijal, U., Uprety, P., & Kovalenko, M. V. (2014). Infrared emitting PbS nanocrystal solids through matrix encapsulation. *Chemistry of Materials*, *26*(14), 4256–4264.

450. Mountrichas, G., Pispas, S., & Tagmatarchis, N. (2007). Aqueous Carbon-Nanotube–Amphiphilic-Block-Copolymer Nanoensembles: Towards Realization of Charge-Transfer Processes with Semiconductor Quantum Dots. *Small*, *3*(3), 404–407.

451. Mountrichas, G., Sandanayaka, A. S., Economopoulos, S. P., Pispas, S., Ito, O., Hasobe, T., & Tagmatarchis, N. (2009). Photoinduced electron transfer in aqueous carbon nanotube/block copolymer/CdS hybrids: application in the construction of photoelectrochemical cells. *Journal of Materials Chemistry*, *19*(47), 8990–8998.

452. Mueller, T., Xia, F., & Avouris, P. (2010). Graphene photodetectors for high-speed optical communications. *Nature Photonics*, *4*(5), 297.

453. Müller-Steinhagen, H., & Trieb, F. (2004). Concentrating solar power. A review of the technology. *Ingenia Inform QR Acad Eng*, *18*, 43–50.

454. Müller, J., Lupton, J. M., Lagoudakis, P. G., Schindler, F., Koeppe, R., Rogach, A. L., & Weller, H. (2005). Wave function engineering in elongated semiconductor nanocrystals with heterogeneous carrier confinement. *Nano Letters*, *5*(10), 2044–2049.

455. Murphy, J. E., Beard, M. C., & Nozik, A. J. (2006). Time-resolved photoconductivity of PbSe nanocrystal arrays. *The Journal of Physical Chemistry B*, *110*(50), 25455–25461.

456. Murphy, J. E., Beard, M. C., Norman, A. G., Ahrenkiel, S. P., Johnson, J. C., Yu, P., & Nozik, A. J. (2006). PbTe colloidal nanocrystals: synthesis, characterization, and multiple exciton generation. *Journal of the American Chemical Society*, *128*(10), 3241–3247.

457. Murray, C., Norris, D. J., & Bawendi, M. G. (1993). Synthesis and characterization of nearly monodisperse CdE (E= sulfur, selenium, tellurium) semiconductor nanocrystallites. *Journal of the American Chemical Society*, *115*(19), 8706–8715.

458. Mutig, A., Blokhin, S. A., Nadtochiy, A. M., Fiol, G., Lott, J. A., Shchukin, V. A., & Bimberg, D. (2009). Frequency response of large aperture oxide-confined 850 nm vertical cavity surface emitting lasers. *Applied Physics Letters*, *95*(13), 131101.

459. Mutig, A., Fiol, G., Moser, P., Arsenijevic, D., Shchukin, V. A., Ledentsov, N. N., & Hopfer, F. (2008). 120 C 20 Gbit/s operation of 980 nm VCSEL. *Electronics Letters*, *44*(22), 1.

460. Na, Y. J., Kim, H. S., & Park, J. (2008). Morphology-controlled lead selenide nanocrystals and their in situ growth on carbon nanotubes. *The Journal of Physical Chemistry C*, *112*(30), 11218–11226.

461. Nadenau, V., Rau, U., Jasenek, A., & Schock, H. W. (2000). Electronic properties of CuGaSe 2-based heterojunction solar cells. Part I. Transport analysis. *Journal of Applied Physics*, *87*(1), 584–593.

462. Nagpal, P., & Klimov, V. I. (2011). Role of mid-gap states in charge transport and photoconductivity in semiconductor nanocrystal films. *Nature Communications*, *2*, 486.

463. Nalwa, H. S. (2004). *Encyclopedia of Nanoscience and Nanotechnology; Volume 7*. American Scientific Publishers.

464. Nam, S., Oh, N., Zhai, Y., & Shim, M. (2015). High efficiency and optical anisotropy in double-heterojunction nanorod light-emitting diodes. *ACS Nano*, *9*(1), 878–885.

465. Nan, W., Niu, Y., Qin, H., Cui, F., Yang, Y., Lai, R., & Peng, X. (2012). Crystal structure control of zinc-blende CdSe/CdS core/shell nanocrystals: synthesis and structure-dependent optical properties. *Journal of the American Chemical Society*, *134*(48), 19685–19693.

466. Neo, D. C., Cheng, C., Stranks, S. D., Fairclough, S. M., Kim, J. S., Kirkland, A. I., & Watt, A. A. (2014). Influence of shell thickness and surface passivation on PbS/CdS core/shell colloidal quantum dot solar cells. *Chemistry of Materials*, *26*(13), 4004–4013.

467. Nesheva, D., Raptis, C., Perakis, A., Bineva, I., Aneva, Z., Levi, Z., & Hofmeister, H. (2002). Raman scattering and photoluminescence from Si nanoparticles in annealed SiO x thin films. *Journal of Applied Physics*, *92*(8), 4678–4683.

468. Neto, A. C., Guinea, F., Peres, N. M., Novoselov, K. S., & Geim, A. K. (2009). The electronic properties of graphene. *Reviews of Modern Physics*, *81*(1), 109.

469. Ning, Z., Dong, H., Zhang, Q., Voznyy, O., & Sargent, E. H. (2014). Solar cells based on inks of n-type colloidal quantum dots. *ACS Nano*, *8*(10), 10321–10327.

470. Ning, Z., Gong, X., Comin, R., Walters, G., Fan, F., Voznyy, O., & Sargent, E. H. (2015). Quantum-dot-in-perovskite solids. *Nature*, *523*(7560), 324.

471. Ning, Z., Voznyy, O., Pan, J., Hoogland, S., Adinolfi, V., Xu, J., & Kemp, K. W. (2014). Air-stable n-type colloidal quantum dot solids. *Nature Materials*, *13*(8), 822.

472. Nootz, G., Padilha, L. A., Levina, L., Sukhovatkin, V., Webster, S., Brzozowski, L., & Van Stryland, E. W. (2011). Size dependence of carrier dynamics and carrier multiplication in PbS quantum dots. *Physical Review B*, *83*(15), 155302.

473. Novoselov, K. S., Geim, A. K., Morozov, S. V., Jiang, D., Zhang, Y., Dubonos, S. V., & Firsov, A. A. (2004). Electric field effect in atomically thin carbon films. *Science*, *306*(5696), 666–669.

474. Novoselov, K. S., Geim, A. K., Morozov, S., Jiang, D., Katsnelson, M., Grigorieva, I., & Firsov, A. A. (2005). Two-dimensional gas of massless Dirac fermions in graphene. *Nature*, *438*(7065), 197.

475. Nozik, A. J. (2001). Spectroscopy and hot electron relaxation dynamics in semiconductor quantum wells and quantum dots. *Annual Review Of Physical Chemistry*, *52*(1), 193–231.

476. Nozik, A. J. (2002). Quantum dot solar cells. *Physica E: Low-dimensional Systems and Nanostructures*, *14*(1–2), 115–120.

477. Nozik, A. J. (2008). Multiple exciton generation in semiconductor quantum dots. *Chemical Physics Letters*, *457*(1–3), 3–11.

478. Nozik, A. J., Beard, M. C., Luther, J. M., Law, M., Ellingson, R. J., & Johnson, J. C. (2010). Semiconductor quantum dots and quantum dot arrays and applications of multiple exciton generation to third-generation photovoltaic solar cells. *Chemical Reviews*, *110*(11), 6873–

6890.

479. NREL, L. K. (2015). Best Research-Cell Efficiencies. *57] MA Green," Corrigendum to Solar cell efficiency tables (version 46) [Prog. Photovolt: Res. Appl*, *23*(805812), 1202–1202.

480. O'regan, B., & Grätzel, M. (1991). A low-cost, high-efficiency solar cell based on dye-sensitized colloidal TiO2 films. *Nature*, *353*(6346), 737.

481. Oh, S. J., Berry, N. E., Choi, J. H., Gaulding, E. A., Lin, H., Paik, T., & Kagan, C. R. (2014). Designing high-performance PbS and PbSe nanocrystal electronic devices through stepwise, post-synthesis, colloidal atomic layer deposition. *Nano Letters*, *14*(3), 1559–1566.

482. Oh, S. J., Berry, N. E., Choi, J. H., Gaulding, E. A., Paik, T., Hong, S. H., & Kagan, C. R. (2013). Stoichiometric control of lead chalcogenide nanocrystal solids to enhance their electronic and optoelectronic device performance. *ACS Nano*, *7*(3), 2413–2421.

483. Ohno, Y., Okamoto, S., Maehashi, K., & Matsumoto, K. (2013). Direct electrical detection of DNA hybridization based on electrolyte-gated graphene field-effect transistor. *Japanese Journal of Applied Physics*, *52*(11R), 110107.

484. Orlanducci, S., Sessa, V., Terranova, M. L., Battiston, G. A., Battiston, S., & Gerbasi, R. (2006). Nanocrystalline TiO2 on single walled carbon nanotube arrays: Towards the assembly of organized C/TiO2 nanosystems. *Carbon*, *44*(13), 2839–2843.

485. Oshima, R., Takata, A., & Okada, Y. (2008). Strain-compensated InAs/GaNAs quantum dots for use in high-efficiency solar cells. *Applied Physics Letters*, *93*(8), 083111.

486. Osovsky, R., Cheskis, D., Kloper, V., Sashchiuk, A., Kroner, M., & Lifshitz, E. (2009). Continuous-wave pumping of multiexciton bands in the photoluminescence spectrum of a single CdTe-CdSe core-shell colloidal quantum dot. *Physical Review Letters*, *102*(19), 197401.

487. Osterwald, C. R. (2003). Standards, calibration and testing of PV modules and solar cells. In *Practical Handbook of Photovoltaics* (pp. 793–816).

488. Özdemir, N., & Yakuphanoglu, F. (2006). The effects of particle size and volume fraction of Al 2 O 3 on electronic thermal conductivity of α-Al_2O_3 particulate reinforced aluminum composites (Al/Al 2 O 3 MMC). *The International Journal of Advanced Manufacturing*

Technology, *29*(3–4), 226–229.

489. Padilha, L. A., Stewart, J. T., Sandberg, R. L., Bae, W. K., Koh, W. K., Pietryga, J. M., & Klimov, V. I. (2013). Aspect ratio dependence of auger recombination and carrier multiplication in PbSe nanorods. *Nano Letters*, *13*(3), 1092–1099.

490. Padilha, L. A., Stewart, J. T., Sandberg, R. L., Bae, W. K., Koh, W. K., Pietryga, J. M., & Klimov, V. I. (2013). Carrier multiplication in semiconductor nanocrystals: influence of size, shape, and composition. *Accounts of Chemical Research*, *46*(6), 1261–1269.

491. Padilha, L. A., Stewart, J. T., Sandberg, R. L., Bae, W. K., Koh, W. K., Pietryga, J. M., & Klimov, V. I. (2013). Aspect ratio dependence of auger recombination and carrier multiplication in PbSe nanorods. *Nano Letters*, *13*(3), 1092–1099.

492. Pagona, G., Sandanayaka, A. S., Hasobe, T., Charalambidis, G., Coutsolelos, A. G., Yudasaka, M., & Tagmatarchis, N. (2008). Characterization and photoelectrochemical properties of nanostructured thin film composed of carbon nanohorns covalently functionalized with porphyrins. *The Journal of Physical Chemistry C*, *112*(40), 15735–15741.

493. Pak, C., Woo, J. Y., Lee, K., Kim, W. D., Yoo, Y., & Lee, D. C. (2012). Extending the limit of low-energy photocatalysis: dye reduction with PbSe/CdSe/CdS core/shell/shell nanocrystals of varying morphologies under infrared irradiation. *The Journal of Physical Chemistry C*, *116*(48), 25407–25414.

494. Pal, B. N., Ghosh, Y., Brovelli, S., Laocharoensuk, R., Klimov, V. I., Hollingsworth, J. A., & Htoon, H. (2011). 'Giant'CdSe/CdS core/shell nanocrystal quantum dots as efficient electroluminescent materials: strong influence of shell thickness on light-emitting diode performance. *Nano Letters*, *12*(1), 331–336.

495. Pan, Z., Mora-Seró, I., Shen, Q., Zhang, H., Li, Y., Zhao, K., & Bisquert, J. (2014). High-efficiency "green" quantum dot solar cells. *Journal of the American Chemical Society*, *136*(25), 9203–9210.

496. Pan, Z., Zhao, K., Wang, J., Zhang, H., Feng, Y., & Zhong, X. (2013). Near infrared absorption of CdSe x Te1–x alloyed quantum dot sensitized solar cells with more than 6% efficiency and high stability. *ACS Nano*, *7*(6), 5215–5222.

497. Pandey, A., & Guyot-Sionnest, P. (2008). Slow electron cooling in

colloidal quantum dots. *Science*, *322*(5903), 929–932.

498. Panjikar, S., Parthasarathy, V., Lamzin, V. S., Weiss, M. S., & Tucker, P. A. (2009). On the combination of molecular replacement and single-wavelength anomalous diffraction phasing for automated structure determination. *Acta Crystallographica Section D: Biological Crystallography*, *65*(10), 1089–1097.
499. Park, S. H., Roy, A., Beaupré, S., Cho, S., Coates, N., Moon, J. S., & Heeger, A. J. (2009). Bulk heterojunction solar cells with internal quantum efficiency approaching 100%. *Nature Photonics*, *3*(5), 297.
500. Park, S. J., Link, S., Miller, W. L., Gesquiere, A., & Barbara, P. F. (2007). Effect of electric field on the photoluminescence intensity of single CdSe nanocrystals. *Chemical Physics*, *341*(1–3), 169–174.
501. Park, Y. S., Ghosh, Y., Chen, Y., Piryatinski, A., Xu, P., Mack, N. H., & Htoon, H. (2013). Super-poissonian statistics of photon emission from single cdse-cds core-shell nanocrystals coupled to metal nanostructures. *Physical Review Letters*, *110*(11), 117401.
502. Parker, D., & Singh, D. J. (2010). High-temperature thermoelectric performance of heavily doped PbSe. *Physical Review B*, *82*(3), 035204.
503. Pasricha, R., Gupta, S., & Srivastava, A. K. (2009). A Facile and Novel Synthesis of Ag–Graphene-Based Nanocomposites. *Small*, *5*(20), 2253–2259.
504. Pattantyus-Abraham, A. G., Kramer, I. J., Barkhouse, A. R., Wang, X., Konstantatos, G., Debnath, R., & Sargent, E. H. (2010). Depleted-heterojunction colloidal quantum dot solar cells. *ACS Nano*, *4*(6), 3374–3380.
505. Pavesi, L., Dal Negro, L., Mazzoleni, C., Franzo, G., & Priolo, F. (2000). Optical gain in silicon nanocrystals. *Nature*, *408*(6811), 440.
506. Pazos-Outón, L. M., Szumilo, M., Lamboll, R., Richter, J. M., Crespo-Quesada, M., Abdi-Jalebi, M., & Ehrler, B. (2016). Photon recycling in lead iodide perovskite solar cells. *Science*, *351*(6280), 1430–1433.
507. Peng, J., Gao, W., Gupta, B. K., Liu, Z., Romero-Aburto, R., Ge, L., & Vithayathil, S. A. (2012). Graphene quantum dots derived from carbon fibers. *Nano Letters*, *12*(2), 844–849.
508. Peng, X. (2010). Band gap and composition engineering on a nanocrystal (BCEN) in solution. *Accounts of Chemical Research*, *43*(11), 1387–1395.
509. Peng, X., Manna, L., Yang, W., Wickham, J., Scher, E., Kadavanich,

A., & Alivisatos, A. P. (2000). Shape control of CdSe nanocrystals. *Nature*, *404*(6773), 59.

510. Perálvarez, M., García, C., López, M., Garrido, B., Barreto, J., Domínguez, C., & Rodríguez, J. A. (2006). Field effect luminescence from Si nanocrystals obtained by plasma-enhanced chemical vapor deposition. *Applied Physics Letters*, *89*(5), 051112.
511. Pi, X., Li, Q., Li, D., & Yang, D. (2011). Spin-coating silicon-quantum-dot ink to improve solar cell efficiency. *Solar Energy Materials and Solar Cells*, *95*(10), 2941–2945.
512. Pijpers, J. J. H., Ulbricht, R., Tielrooij, K. J., Osherov, A., Golan, Y., Delerue, C., Bonn, M., et al. (2009). Assessment of carrier-multiplication efficiency in bulk PbSe and PbS. *Nature Physics*, *5*(11), 811.
513. Piva, P. G., Goldberg, R. D., Mitchell, I. V., Labrie, D., Leon, R., Charbonneau, S., & Fafard, S. (2000). Enhanced degradation resistance of quantum dot lasers to radiation damage. *Applied Physics Letters*, *77*(5), 624–626.
514. Polman, A., & Atwater, H. A. (2012). Photonic design principles for ultrahigh-efficiency photovoltaics. *Nature Materials*, *11*(3), 174.
515. Popescu, V., Bester, G., Hanna, M. C., Norman, A. G., & Zunger, A. (2008). Theoretical and experimental examination of the intermediate-band concept for strain-balanced (In, Ga) As/Ga (As, P) quantum dot solar cells. *Physical Review B*, *78*(20), 205321.
516. Porter, V. J., Geyer, S., Halpert, J. E., Kastner, M. A., & Bawendi, M. G. (2008). Photoconduction in annealed and chemically treated CdSe/ZnS inorganic nanocrystal films. *The Journal of Physical Chemistry C*, *112*(7), 2308–2316.
517. Pourret, A., Guyot-Sionnest, P., & Elam, J. W. (2009). Atomic layer deposition of ZnO in quantum dot thin films. *Advanced Materials*, *21*(2), 232–235.
518. Prasannan, A. and Imae, T. (2013). One-Pot Synthesis of Fluorescent Carbon Dots from Orange Waste Peels. *Industrial & Engineering Chemistry Research, 52*(44), 5673–15678.
519. Prezioso, S., Anopchenko, A., Gaburro, Z., Pavesi, L., Pucker, G., Vanzetti, L., & Bellutti, P. (2008). Electrical conduction and electroluminescence in nanocrystalline silicon-based light emitting devices. *Journal of Applied Physics*, *104*(6), 063103.

520. Prezioso, S., Hossain, S. M., Anopchenko, A., Pavesi, L., Wang, M., Pucker, G., & Bellutti, P. (2009). Superlinear photovoltaic effect in Si nanocrystals-based metal-insulator-semiconductor devices. *Applied Physics Letters*, *94*(6), 062108.

521. Price, M. B., Butkus, J., Jellicoe, T. C., Sadhanala, A., Briane, A., Halpert, J. E., & Deschler, F. (2015). Hot-carrier cooling and photoinduced refractive index changes in organic–inorganic lead halide perovskites. *Nature Communications*, *6*, 8420.

522. Prokes, S. M., Carlos, W. E., Veprek, S., & Ossadnik, C. (1998). Defect studies in as-deposited and processed nanocrystalline S i/S i O 2 structures. *Physical Review B*, *58*(23), 15632.

523. Pu, C., Zhou, J., Lai, R., Niu, Y., Nan, W., & Peng, X. (2013). Highly reactive, flexible yet green Se precursor for metal selenide nanocrystals: Se-octadecene suspension (Se-SUS). *Nano Research*, *6*(9), 652–670.

524. Pucker, G., Bellutti, P., Spinella, C., Gatterer, K., Cazzanelli, M., & Pavesi, L. (2000). Room temperature luminescence from (Si/SiO 2) n (n= 1, 2, 3) multilayers grown in an industrial low-pressure chemical vapor deposition reactor. *Journal of Applied Physics*, *88*(10), 6044–6051.

525. Pucker, G., Serra, E., & Jestin, Y. (2012). Silicon quantum dots for photovoltaics: a review. In *Quantum Dots-A Variety of New Applications*. InTech.

526. Qian, L., Zheng, Y., Xue, J., & Holloway, P. H. (2011). Stable and efficient quantum-dot light-emitting diodes based on solution-processed multilayer structures. *Nature Photonics*, *5*(9), 543.

527. Qu, K., Wang, J., Ren, J., & Qu, X. (2013). Carbon dots prepared by hydrothermal treatment of dopamine as an effective fluorescent sensing platform for the label-free detection of iron (III) ions and dopamine. *Chemistry-A European Journal, 19*(22), 7243–7249.

528. Querner, C., Reiss, P., Sadki, S., Zagorska, M., & Pron, A. (2005). Size and ligand effects on the electrochemical and spectroelectrochemical responses of CdSe nanocrystals. *Physical Chemistry Chemical Physics*, *7*(17), 3204–3209.

529. Rabani, E., & Baer, R. (2010). Theory of multiexciton generation in semiconductor nanocrystals. *Chemical Physics Letters*, *496*(4–6), 227–235.

530. Rahim, R. A., Kurahashi, H., Uesugi, K., & Fukuda, H. (2007). Carbon

nano dots scale by focused ion beam system for MIS diode nano devices. *Surface Science*, *601*(22), 5112–5115.

531. Rauf, I. A., & Rezai, P. (2017). A review of materials selection for optimized efficiency in quantum dot sensitized solar cells: a simplified approach to reviewing literature data. *Renewable and Sustainable Energy Reviews*, *73*, 408–422.
532. Ray, S. C., Saha, A., Jana, N. R., & Sarkar, R. (2009). Fluorescent carbon nanoparticles: synthesis, characterization, and bioimaging application. *The Journal of Physical Chemistry C*, *113*(43), 18546–18551.
533. Recommendation, I. T. U. (2001). The control of jitter and wander within the optical transport network (OTN). *Recommendation G, 8251.*
534. Regulacio, M. D., & Han, M. Y. (2010). Composition-tunable alloyed semiconductor nanocrystals. *Accounts of Chemical Research*, *43*(5), 621–630.
535. Rhee, J. H., Chung, C. C., & Diau, E. W. G. (2013). A perspective of mesoscopic solar cells based on metal chalcogenide quantum dots and organometal-halide perovskites. *NPG Asia Materials*, *5*(10), e68.
536. Ribbat, C., Sellin, R. L., Kaiander, I., Hopfer, F., Ledentsov, N. N., Bimberg, D., & Maximov, M. V. (2003). Complete suppression of filamentation and superior beam quality in quantum-dot lasers. *Applied Physics Letters*, *82*(6), 952–954.
537. Ribbat, C., Sellin, R., Grundmann, M., Bimberg, D., Sobolev, N. A., & Carmo, M. C. (2001). Enhanced radiation hardness of quantum dot lasers to high energy proton irradiation. *Electronics Letters*, *37*(3), 174–175.
538. Richards, B. S. (2006). Enhancing the performance of silicon solar cells via the application of passive luminescence conversion layers. *Solar Energy Materials and Solar Cells*, *90*(15), 2329–2337.
539. Richards, B. S. (2006). Luminescent layers for enhanced silicon solar cell performance: Down-conversion. *Solar Energy Materials and Solar Cells*, *90*(9), 1189–1207.
540. Richards, B. S., & McIntosh, K. R. (2007). Overcoming the poor short wavelength spectral response of CdS/CdTe photovoltaic modules via luminescence down-shifting: ray-tracing simulations. *Progress in Photovoltaics: Research and Applications*, *15*(1), 27–34.
541. Rinnert, H., & Vergnat, M. (2003). Influence of the temperature on

the photoluminescence of silicon clusters embedded in a silicon oxide matrix. *Physica E: Low-dimensional Systems and Nanostructures*, *16*(3–4), 382–387.

542. Robbins, D. J. (1980). Aspects of the theory of impact ionization in semiconductors (I). *Physica Status Solidi (b)*, *97*(1), 9–50.
543. Robel, I., Gresback, R., Kortshagen, U., Schaller, R. D., & Klimov, V. I. (2009). Universal size-dependent trend in Auger recombination in direct-gap and indirect-gap semiconductor nanocrystals. *Physical Review Letters*, *102*(17), 177404.
544. Robel, I., Kuno, M., & Kamat, P. V. (2007). Size-dependent electron injection from excited CdSe quantum dots into TiO2 nanoparticles. *Journal of the American Chemical Society*, *129*(14), 4136–4137.
545. Robel, I., Subramanian, V., Kuno, M., & Kamat, P. V. (2006). Quantum dot solar cells. Harvesting light energy with CdSe nanocrystals molecularly linked to mesoscopic TiO2 films. *Journal of the American Chemical Society*, *128*(7), 2385–2393.
546. Ross, R. T., & Nozik, A. J. (1982). Efficiency of hot-carrier solar energy converters. *Journal of Applied Physics*, *53*(5), 3813–3818.
547. Rothenberg, E., Kazes, M., Shaviv, E., & Banin, U. (2005). Electric field induced switching of the fluorescence of single semiconductor quantum rods. *Nano Letters*, *5*(8), 1581–1586.
548. Ruan, J., Fauchet, P. M., Dal Negro, L., Cazzanelli, M., & Pavesi, L. (2003). Stimulated emission in nanocrystalline silicon superlattices. *Applied Physics Letters*, *83*(26), 5479–5481.
549. Rühle, S., Shalom, M., & Zaban, A. (2010). Quantum-dot-sensitized solar cells. *ChemPhysChem*, *11*(11), 2290–2304.
550. Ruiz-Rosas, R., Bedia, J., Lallave, M., Loscertales, I., Barrero, A., Rodríguez-Mirasol, J. and Cordero, T. (2010). The production of submicron diameter carbon fibers by the electrospinning of lignin. *Carbon, 48*(3), 696–705.
551. Rupasov, V. I., & Klimov, V. I. (2007). Carrier multiplication in semiconductor nanocrystals via intraband optical transitions involving virtual biexciton states. *Physical Review B*, *76*(12), 125321.
552. Sahu, S., Behera, B., Maiti, T. K., & Mohapatra, S. (2012). Simple one-step synthesis of highly luminescent carbon dots from orange juice: application as excellent bio-imaging agents. *Chemical Communications*, *48*(70), 8835–8837.

553. Saito, H., Nishi, K., Ogura, I., Sugou, S., & Sugimoto, Y. (1996). Room-temperature lasing operation of a quantum-dot vertical-cavity surface-emitting laser. *Applied Physics Letters*, *69*(21), 3140–3142.

554. Sambur, J. B., & Parkinson, B. A. (2010). CdSe/ZnS core/shell quantum dot sensitization of low index TiO_2 single crystal surfaces. *Journal of the American Chemical Society*, *132*(7), 2130–2131.

555. Sambur, J. B., Novet, T., & Parkinson, B. A. (2010). Multiple exciton collection in a sensitized photovoltaic system. *Science*, *330*(6000), 63–66.

556. Sambur, J. B., Riha, S. C., Choi, D., & Parkinson, B. A. (2010). Influence of surface chemistry on the binding and electronic coupling of CdSe quantum dots to single crystal TiO_2 surfaces. *Langmuir*, *26*(7), 4839–4847.

557. Sandberg, R. L., Padilha, L. A., Qazilbash, M. M., Bae, W. K., Schaller, R. D., Pietryga, J. M., & Klimov, V. I. (2012). Multiexciton dynamics in infrared-emitting colloidal nanostructures probed by a superconducting nanowire single-photon detector. *ACS Nano*, *6*(11), 9532–9540.

558. Sandeep, C. S., Ten Cate, S., Schins, J. M., Savenije, T. J., Liu, Y., Law, M., & Siebbeles, L. D. (2013). High charge-carrier mobility enables exploitation of carrier multiplication in quantum-dot films. *Nature Communications*, *4*, 2360.

559. Santra, P. K., & Kamat, P. V. (2012). Mn-doped quantum dot sensitized solar cells: a strategy to boost efficiency over 5%. *Journal of the American Chemical Society*, *134*(5), 2508–2511.

560. Santra, P. K., & Kamat, P. V. (2013). Tandem-layered quantum dot solar cells: tuning the photovoltaic response with luminescent ternary cadmium chalcogenides. *Journal of the American Chemical Society*, *135*(2), 877–885.

561. Sargent, E. H. (2012). Colloidal quantum dot solar cells. *Nature Photoanics*, *6*(3), 133.

562. Sayevich, V., Gaponik, N., Plötner, M., Kruszynska, M., Gemming, T., Dzhagan, V. M., & Eychmüller, A. (2015). Stable dispersion of iodide-capped PbSe quantum dots for high-performance low-temperature processed electronics and optoelectronics. *Chemistry of Materials*, *27*(12), 4328–4337.

563. Schaller, R. D., & Klimov, V. I. (2004). High efficiency carrier multiplication in PbSe nanocrystals: implications for solar energy

conversion. *Physical Review Letters*, *92*(18), 186601.

564. Schaller, R. D., Agranovich, V. M., & Klimov, V. I. (2005). High-efficiency carrier multiplication through direct photogeneration of multi-excitons via virtual single-exciton states. *Nature Physics*, *1*(3), 189.

565. Schaller, R. D., Petruska, M. A., & Klimov, V. I. (2005). Effect of electronic structure on carrier multiplication efficiency: Comparative study of PbSe and CdSe nanocrystals. *Applied Physics Letters*, *87*(25), 253102.

566. Schaller, R. D., Sykora, M., Jeong, S., & Klimov, V. I. (2006). High-efficiency carrier multiplication and ultrafast charge separation in semiconductor nanocrystals studied via time-resolved photoluminescence. *The Journal of Physical Chemistry B*, *110*(50), 25332–25338.

567. Schapotschnikow, P., van Huis, M. A., Zandbergen, H. W., Vanmaekelbergh, D., & Vlugt, T. J. (2010). Morphological transformations and fusion of PbSe nanocrystals studied using atomistic simulations. *Nano letters*, *10*(10), 3966–3971.

568. Schliehe, C., Juarez, B. H., Pelletier, M., Jander, S., Greshnykh, D., Nagel, M., & Weller, H. (2010). Ultrathin PbS sheets by two-dimensional oriented attachment. *Science*, *329*(5991), 550–553.

569. Schmeckebier, H., Fiol, G., Meuer, C., Arsenijević, D., & Bimberg, D. (2010). Complete pulse characterization of quantum-dot mode-locked lasers suitable for optical communication up to 160 Gbit/s. *Optics Express*, *18*(4), 3415–3425.

570. Schmeckebier, H., Meuer, C., Bimberg, D., Schmidt-Langhorst, C., Galperin, A., & Schubert, C. (2010). Quantum dot semiconductor optical amplifiers at 1.3 μm for applications in all-optical communication networks. *Semiconductor Science and Technology*, *26*(1), 014009.

571. Schmidbaur, H. and Schier, A. (2012). Coordination Chemistry at Carbon: The Patchwork Family Comprising (Ph3P)2C, (Ph3P)C(C2H4), and (C2H4)2C. *Angewandte Chemie International Edition*, *52*(1), 176–186.

572. Schmidt, O. G., Kirstaedter, N., Ledentsov, N. N., Mao, M. H., Bimberg, D., Ustinov, V. M., & Alferov, Z. I. (1996). Prevention of gain saturation by multi-layer quantum dot lasers. *Electronics Letters*, *32*(14), 1302–1304.

573. Schneider, S., Borri, P., Langbein, W., Woggon, U., Sellin, R. L., Ouyang, D., & Bimberg, D. (2005). Excited-state gain dynamics in InGaAs quantum-dot amplifiers. *IEEE Photonics Technology Letters*, *17*(10), 2014–2016.

574. Schulze, T. F., & Schmidt, T. W. (2015). Photochemical upconversion: present status and prospects for its application to solar energy conversion. *Energy & Environmental Science*, *8*(1), 103–125.

575. Schur, R., Sogawa, F., & Nishioka, M. (1997). S. Ishida, Y. Arakawa. *Jpn. J Appl. Phys*, *35*, L357.

576. Selinsky, R. S., Ding, Q., Faber, M. S., Wright, J. C., & Jin, S. (2013). Quantum dot nanoscale heterostructures for solar energy conversion. *Chemical Society Reviews*, *42*(7), 2963–2985.

577. Selinsky, R. S., Shin, S., Lukowski, M. A., & Jin, S. (2012). Epitaxial Heterostructures of Lead Selenide Quantum Dots on Hematite Nanowires. *The Journal of Physical Chemistry Letters*, *3*(12), 1649–1656.

578. Semonin, O. E., Luther, J. M., & Beard, M. C. (2012). Quantum dots for next-generation photovoltaics. *Materials Today*, *15*(11), 508–515.

579. Semonin, O. E., Luther, J. M., Choi, S., Chen, H. Y., Gao, J., Nozik, A. J., & Beard, M. C. (2011). Peak external photocurrent quantum efficiency exceeding 100% via MEG in a quantum dot solar cell. *Science*, *334*(6062), 1530–1533.

580. Sensfuss, S., & Al-Ibrahim, M. (2005). Optoelectronic properties of conjugated polymer/fullerene binary pairs with variety of LUMO level differences. *Optical Engineering.* New York-Marcel Dekker Incorporated, *99*, 529.

581. Service, R. F. (1996). new solar cells seem to hawe power at the right price. *Science (Washington, D. C.)*, *272*(5269).

582. Shabaev, A., Efros, A. L., & Nozik, A. J. (2006). Multiexciton generation by a single photon in nanocrystals. *Nano Letters*, *6*(12), 2856–2863.

583. Shaklee, K. Á., & Nahory, R. E. (1970). Valley-orbit splitting of free excitons? The absorption edge of Si. *Physical Review Letters*, *24*(17), 942.

584. Shalom, M., Albero, J., Tachan, Z., Martínez-Ferrero, E., Zaban, A., & Palomares, E. (2010). Quantum dot− dye bilayer-sensitized solar cells: breaking the limits imposed by the low absorbance of dye monolayers. *The Journal of Physical Chemistry Letters*, *1*(7), 1134–1138.

585. Shchukin, V. A., Ledentsov, N. N., Lott, J. A., Quast, H., Hopfer, F., Karachinsky, L. Y., & Kalosha, V. P. (2008, February). Ultra high-speed electro-optically modulated VCSELs: modeling and experimental results. In *Physics and Simulation of Optoelectronic Devices XVI* (Vol. 6889, p. 68890H). International Society for Optics and Photonics.

586. Shen, Q., Yamada, A., Tamura, S., & Toyoda, T. (2010). CdSe quantum dot-sensitized solar cell employing TiO 2 nanotube working-electrode and Cu 2 S counter-electrode. *Applied Physics Letters*, *97*(12), 123107.

587. Shi, Y., Dong, X., Chen, P., Wang, J., & Li, L. J. (2009). Effective doping of single-layer graphene from underlying SiO 2 substrates. *Physical Review B*, *79*(11), 115402.

588. Shi, Y., Fang, W., Zhang, K., Zhang, W., & Li, L. J. (2009). Photoelectrical Response in Single-Layer Graphene Transistors. *Small*, *5*(17), 2005–2011.

589. Shim, H. C., Jeong, S., & Han, C. S. (2011). Controlled assembly of CdSe/MWNT hybrid material and its fast photoresponse with wavelength selectivity. *Nanotechnology*, *22*(16), 165201.

590. Shirasaki, Y., Supran, G. J., Bawendi, M. G., & Bulović, V. (2013). Emergence of colloidal quantum-dot light-emitting technologies. *Nature Photonics*, *7*(1), 13.

591. Shirasaki, Y., Supran, G. J., Tisdale, W. A., & Bulović, V. (2013). Origin of efficiency roll-off in colloidal quantum-dot light-emitting diodes. *Physical Review Letters*, *110*(21), 217403.

592. Shockley, W., & Queisser, H. J. (1961). Detailed balance limit of efficiency of p-n junction solar cells. *Journal of Applied Physics*, *32*(3), 510–519.

593. Shrestha, S. K., Aliberti, P., & Conibeer, G. J. (2010). Energy selective contacts for hot carrier solar cells. *Solar Energy Materials and Solar Cells*, *94*(9), 1546–1550.

594. Šimurda, M., Němec, P., Formanek, P., Němec, I., Němcová, Y., & Malý, P. (2006). Morphology of CdSe films prepared by chemical bath deposition: The role of substrate. *Thin Solid Films*, *511*, 71–75.

595. Sinke, W. C., Ballif, C., Bett, A., Dimmler, B., Dimova-Malinovska, D., Fath, P., & Martí, A. (2007). *A Strategic Research Agenda for Photovoltaic Solar Energy Technology* (No. PV-LAB-CONF-2007–004).

596. Sites, J. R., Tavakolian, H., & Sasala, R. A. (1990). Analysis of apparent

quantum efficiency. *Solar Cells*, *29*(1), 39–48.

597. Sitt, A., Hadar, I., & Banin, U. (2013). Band-gap engineering, optoelectronic properties and applications of colloidal heterostructured semiconductor nanorods. *Nano Today*, *8*(5), 494–513.

598. Smith, C., & Binks, D. (2013). Multiple exciton generation in colloidal nanocrystals. *Nanomaterials*, *4*(1), 19–45.

599. Smith, M. B., & Michl, J. (2010). Singlet fission. *Chemical Reviews*, *110*(11), 6891–6936.

600. Smith, M. B., & Michl, J. (2013). Recent advances in singlet fission. *Annual Review of Physical Chemistry*, *64*, 361–386.

601. Smowton, P. M., Pearce, E. J., Schneider, H. C., Chow, W. W., & Hopkinson, M. (2002). Filamentation and linewidth enhancement factor in InGaAs quantum dot lasers. *Applied Physics Letters*, *81*(17), 3251–3253.

602. Snaith, H. J. (2012). How should you measure your excitonic solar cells? *Energy & Environmental Science*, *5*(4), 6513–6520.

603. Snaith, H. J. (2012). The perils of solar cell efficiency measurements. *Nature Photonics*, *6*(6), 337.

604. Solar Generation, V. (2008). Solar electricity for over 1 billion people and 2 million jobs by 2020. European photovoltaic industry association, available online at http://www. epia.org/publications/epia-publications. html [accessed Sep. 2010].

605. Song, D., Cho, E. C., Conibeer, G., Flynn, C., Huang, Y., & Green, M. A. (2008). Structural, electrical and photovoltaic characterization of Si nanocrystals embedded SiC matrix and Si nanocrystals/c-Si heterojunction devices. *Solar Energy Materials and Solar Cells*, *92*(4), 474–481.

606. Speirs, M. J., Balazs, D. M., Fang, H. H., Lai, L. H., Protesescu, L., Kovalenko, M. V., & Loi, M. A. (2015). Origin of the increased open circuit voltage in PbS–CdS core–shell quantum dot solar cells. *Journal of Materials Chemistry A*, *3*(4), 1450–1457.

607. Spinicelli, P., Buil, S., Quelin, X., Mahler, B., Dubertret, B., & Hermier, J. P. (2009). Bright and grey states in CdSe-CdS nanocrystals exhibiting strongly reduced blinking. *Physical Review Letters*, *102*(13), 136801.

608. Steckel, J. S., Snee, P., Coe-Sullivan, S., Zimmer, J. P., Halpert, J. E., Anikeeva, P., & Bawendi, M. G. (2006). Color-Saturated Green-Emitting QD-LEDs. *Angewandte Chemie International Edition*,

45(35), 5796–5799.

609. Stegner, A. R., Pereira, R. N., Klein, K., Lechner, R., Dietmueller, R., Brandt, M. S., & Wiggers, H. (2008). Electronic transport in phosphorus-doped silicon nanocrystal networks. *Physical Review Letters*, *100*(2), 026803.

610. Stewart, J. T., Padilha, L. A., Bae, W. K., Koh, W. K., Pietryga, J. M., & Klimov, V. I. (2013). Carrier multiplication in quantum dots within the framework of 2 competing energy relaxation mechanisms. *The Journal of Physical Chemistry Letters*, *4*(12), 2061–2068.

611. Stier, O., Grundmann, M., & Bimberg, D. (1999). Electronic and optical properties of strained quantum dots modeled by 8-band k□ p theory. *Physical Review B*, *59*(8), 5688.

612. Stupca, M., Alsalhi, M., Al Saud, T., Almuhanna, A., & Nayfeh, M. H. (2007). Enhancement of polycrystalline silicon solar cells using ultrathin films of silicon nanoparticle. *Applied Physics Letters*, *91*(6), 063107.

613. Sukhovatkin, V., Hinds, S., Brzozowski, L., & Sargent, E. H. (2009). Colloidal quantum-dot photodetectors exploiting multiexciton generation. *Science*, *324*(5934), 1542–1544.

614. Sun, B., & Greenham, N. C. (2006). Improved efficiency of photovoltaics based on CdSe nanorods and poly (3-hexylthiophene) nanofibers. *Physical Chemistry Chemical Physics*, *8*(30), 3557–3560.

615. Sun, B., Findikoglu, A. T., Sykora, M., Werder, D. J., & Klimov, V. I. (2009). Hybrid photovoltaics based on semiconductor nanocrystals and amorphous silicon. *Nano Letters*, *9*(3), 1235–1241.

616. Sun, B., Snaith, H. J., Dhoot, A. S., Westenhoff, S., & Greenham, N. C. (2005). Vertically segregated hybrid blends for photovoltaic devices with improved efficiency. *Journal of Applied Physics*, *97*(1), 014914.

617. Sun, M., Ma, X., Chen, X., Sun, Y., Cui, X., & Lin, Y. (2014). A nanocomposite of carbon quantum dots and TiO 2 nanotube arrays: enhancing photoelectrochemical and photocatalytic properties. *RSC Advances*, *4*(3), 1120–1127.

618. Sun, X. W., Chen, J., Song, J. L., Zhao, D. W., Deng, W. Q., & Lei, W. (2010). Ligand capping effect for dye solar cells with a CdSe quantum dot sensitized ZnO nanorod photoanode. *Optics Express*, *18*(2), 1296–1301.

619. Sun, Y. P., Wang, X., Lu, F., Cao, L., Meziani, M. J., Luo, P. G., &

Veca, L. M. (2008). Doped carbon nanoparticles as a new platform for highly photoluminescent dots. *The Journal of Physical Chemistry C, 112*(47), 18295–18298.

620. Sun, Y. P., Zhou, B., Lin, Y., Wang, W., Fernando, K. S., Pathak, P., & Luo, P. G. (2006). Quantum-sized carbon dots for bright and colorful photoluminescence. *Journal of the American Chemical Society, 128*(24), 7756–7757.

621. Sun, Z., Liu, Z., Li, J., Tai, G. A., Lau, S. P., & Yan, F. (2012). Infrared photodetectors based on CVD-grown graphene and PbS quantum dots with ultrahigh responsivity. *Advanced Materials*, *24*(43), 5878–5883.

622. Supran, G. J., Shirasaki, Y., Song, K. W., Caruge, J. M., Kazlas, P. T., Coe-Sullivan, S., & Bulović, V. (2013). QLEDs for displays and solid-state lighting. *Mrs Bulletin*, *38*(9), 703–711.

623. Supran, G. J., Song, K. W., Hwang, G. W., Correa, R. E., Scherer, J., Dauler, E. A., & Bulović, V. (2015). High-performance shortwave-infrared light-emitting devices using core–shell (PbS–CdS) colloidal quantum dots. *Advanced Materials*, *27*(8), 1437–1442.

624. Švrček, V., Mariotti, D., Shibata, Y., & Kondo, M. (2010). A hybrid heterojunction based on fullerenes and surfactant-free, self-assembled, closely packed silicon nanocrystals. *Journal of Physics D: Applied Physics*, *43*(41), 415402.

625. Švrček, V., Slaoui, A., & Muller, J. C. (2004). Silicon nanocrystals as light converter for solar cells. *Thin Solid Films*, *451*, 384–388.

626. Swathi, R. S., & Sebastian, K. L. (2009). Long range resonance energy transfer from a dye molecule to graphene has (distance)−4 dependence. *The Journal of Chemical Physics*, *130*(8), 086101.

627. Sychugov, I., Juhasz, R., Valenta, J., & Linnros, J. (2005). Narrow luminescence linewidth of a silicon quantum dot. *Physical Review Letters*, *94*(8), 087405.

628. Tabachnyk, M., Ehrler, B., Bayliss, S., Friend, R. H., & Greenham, N. C. (2013). Triplet diffusion in singlet exciton fission sensitized pentacene solar cells. *Applied Physics Letters*, *103*(15), 190–191.

629. Tabachnyk, M., Ehrler, B., Gélinas, S., Böhm, M. L., Walker, B. J., Musselman, K. P., & Rao, A. (2014). Resonant energy transfer of triplet excitons from pentacene to PbSe nanocrystals. *Nature Materials*, *13*(11), 1033.

630. Tabachnyk, M., Karani, A. H., Broch, K., Pazos-Outón, L. M., Xiao,

J., Jellicoe, T. C., & Greenham, N. C. (2016). Efficient singlet exciton fission in pentacene prepared from a soluble precursor. *APL Materials*, *4*(11), 116112.

631. Takeda, Y., Ito, T., Motohiro, T., König, D., Shrestha, S., & Conibeer, G. (2009). Hot carrier solar cells operating under practical conditions. *Journal of Applied Physics*, *105*(7), 074905.
632. Talapin, D. V., & Murray, C. B. (2005). PbSe nanocrystal solids for n-and p-channel thin film field-effect transistors. *Science*, *310*(5745), 86–89.
633. Talapin, D. V., Lee, J. S., Kovalenko, M. V., & Shevchenko, E. V. (2009). Prospects of colloidal nanocrystals for electronic and optoelectronic applications. *Chemical Reviews*, *110*(1), 389–458.
634. Talgorn, E., Gao, Y., Aerts, M., Kunneman, L. T., Schins, J. M., Savenije, T. J., & Siebbeles, L. D. (2011). Unity quantum yield of photogenerated charges and band-like transport in quantum-dot solids. *Nature Nanotechnology*, *6*(11), 733.
635. Tamang, A., Hongsingthong, A., Sichanugrist, P., Jovanov, V., Gebrewold, H. T., Konagai, M., & Knipp, D. (2016). On the potential of light trapping in multiscale textured thin film solar cells. *Solar Energy Materials and Solar Cells*, *144*, 300–308.
636. Tang, J., & Sargent, E. H. (2011). Infrared colloidal quantum dots for photovoltaics: fundamentals and recent progress. *Advanced Materials*, *23*(1), 12–29.
637. Tang, J., Kemp, K. W., Hoogland, S., Jeong, K. S., Liu, H., Levina, L., & Chou, K. W. (2011). Colloidal-quantum-dot photovoltaics using atomic-ligand passivation. *Nature Materials*, *10*(10), 765.
638. Tang, J., Liu, H., Zhitomirsky, D., Hoogland, S., Wang, X., Furukawa, M., & Sargent, E. H. (2012). Quantum junction solar cells. *Nano Letters*, *12*(9), 4889–4894.
639. Tang, J., Wang, X., Brzozowski, L., Barkhouse, D. A. R., Debnath, R., Levina, L., & Sargent, E. H. (2010). Schottky quantum dot solar cells stable in air under solar illumination. *Advanced Materials*, *22*(12), 1398–1402.
640. Tang, Y. B., Lee, C. S., Xu, J., Liu, Z. T., Chen, Z. H., He, Z., & Luo, L. (2010). Incorporation of graphenes in nanostructured TiO2 films via molecular grafting for dye-sensitized solar cell application. *Acs Nano*, *4*(6), 3482–3488.

641. Tani, Y., Kobayashi, S., & Kawazoe, H. (2008). Ion beam deposition of quantum dots from colloidal solution. *Japanese Journal of Applied Physics*, *47*(4S), 2977.

642. Tavallali, H., Vahdati, P., & Shaabanpur, E. (2011). Developing a new method of 4-(2-pyridylazo)-resorcinol immobilization on triacetylcellulose membrane for selective determination of Ga3+ in water samples. *Sensors and Actuators B: Chemical*, *159*(1), 154–158.

643. Ten Cate, S., Sandeep, C. S., Liu, Y., Law, M., Kinge, S., Houtepen, A. J., & Siebbeles, L. D. (2015). Generating free charges by carrier multiplication in quantum dots for highly efficient photovoltaics. *Accounts of Chemical Research*, *48*(2), 174–181.

644. Tetsuka, H., Asahi, R., Nagoya, A., Okamoto, K., Tajima, I., Ohta, R., & Okamoto, A. (2012). Optically tunable amino-functionalized graphene quantum dots. *Advanced Materials*, *24*(39), 5333–5338.

645. Thakar, R., Chen, Y., & Snee, P. T. (2007). Efficient emission from core/ (doped) shell nanoparticles: applications for chemical sensing. *Nano Letters*, *7*(11), 3429–3432.

646. Thompson, M. G., Marinelli, C., Zhao, X., Sellin, R. L., Penty, R. V., White, I. H., & Blamire, M. G. (2005). Colliding-pulse modelocked quantum dot lasers. *Electronics Letters*, *41*(5), 248–250.

647. Thompson, N. J., Wilson, M. W., Congreve, D. N., Brown, P. R., Scherer, J. M., Bischof, T. S., & Bulović, V. (2014). Energy harvesting of non-emissive triplet excitons in tetracene by emissive PbS nanocrystals. *Nature Materials*, *13*(11), 1039.

648. Thomson, J. W., Nagashima, K., Macdonald, P. M., & Ozin, G. A. (2011). From sulfur– amine solutions to metal sulfide nanocrystals: peering into the oleylamine– sulfur black box. *Journal of the American Chemical Society*, *133*(13), 5036–5041.

649. Thon, S. M., Ip, A. H., Hoogland, S., Voznyy, O., Zhitomirsky, D., Debnath, R., & Kemp, K. W. (2013, March). Hybrid passivated colloidal quantum dot solids for photovoltaics. In *APS Meeting Abstracts*.

650. Tian, J., Leng, Y., Zhao, Z., Xia, Y., Sang, Y., Hao, P., & Liu, H. (2015). Carbon quantum dots/hydrogenated TiO2 nanobelt heterostructures and their broad spectrum photocatalytic properties under UV, visible, and near-infrared irradiation. *Nano Energy*, *11*, 419–427.

651. Titirici, M., Antonietti, M. and Baccile, N. (2008). Hydrothermal carbon from biomass: a comparison of the local structure from poly-

to monosaccharides and pentoses/hexoses. *Green Chemistry, 10*(11), 1204–1212.

652. Torrisi, F., Hasan, T., Wu, W., Sun, Z., Lombardo, A., Kulmala, T. S., & Chu, D. (2012). Inkjet-printed graphene electronics. *ACS Nano*, *6*(4), 2992–3006.
653. Toyoda, T., Kobayashi, J., & Shen, Q. (2008). Correlation between crystal growth and photosensitization of nanostructured TiO2 electrodes using supporting Ti substrates by self-assembled CdSe quantum dots. *Thin Solid Films*, *516*(9), 2426–2431.
654. Toyoda, T., Uehata, T., Suganuma, R., Tamura, S., Sato, A., Yamamoto, K., Kobayashi, N., et al. (2007). Crystal growth of CdSe quantum dots adsorbed on nanoparticle, inverse opal, and nanotube TiO2 photoelectrodes characterized by photoacoustic spectroscopy. *Japanese Journal of Applied Physics*, *46*(7S), 4616.
655. Trinh, M. T., Limpens, R., De Boer, W. D., Schins, J. M., Siebbeles, L. D., & Gregorkiewicz, T. (2012). Direct generation of multiple excitons in adjacent silicon nanocrystals revealed by induced absorption. *Nature Photonics*, *6*(5), 316.
656. Trinh, M. T., Polak, L., Schins, J. M., Houtepen, A. J., Vaxenburg, R., Maikov, G. I., & Nozik, A. J. (2011). Anomalous independence of multiple exciton generation on different group IV− VI quantum dot architectures. *Nano Letters*, *11*(4), 1623–1629.
657. Trupke, T., Green, M. A., & Würfel, P. (2002). Improving solar cell efficiencies by down-conversion of high-energy photons. *Journal of Applied Physics*, *92*(3), 1668–1674.
658. Tsakalakos, L., Balch, J., Fronheiser, J., Korevaar, B. A., Sulima, O., & Rand, J. (2007). Silicon nanowire solar cells. *Applied Physics Letters*, *91*(23), 233117.
659. Tsang, W. T., Weisbuch, C., Miller, R. C., & Dingle, R. (1979). Current injection GaAs-Al x Ga1− x as multi-quantum-well heterostructure lasers prepared by molecular beam epitaxy. *Applied Physics Letters*, *35*(9), 673–675.
660. Tsujimoto, K., Nguyen, D. C., Ito, S., Nishino, H., Matsuyoshi, H., Konno, A., & Tennakone, K. (2012). TiO2 surface treatment effects by Mg2+, Ba2+, and Al3+ on Sb2S3 extremely thin absorber solar cells. *The Journal of Physical Chemistry C*, *116*(25), 13465–13471.
661. Tsybeskov, L., Vandyshev, J. V., & Fauchet, P. M. (1994). Blue emission

in porous silicon: Oxygen-related photoluminescence. *Physical Review B*, *49*(11), 7821.

662. Turyanska, L., Elfurawi, U., Li, M., Fay, M. W., Thomas, N. R., Mann, S., & Patane, A. (2009). Tailoring the physical properties of thiol-capped PbS quantum dots by thermal annealing. *Nanotechnology*, *20*(31), 315604.

663. Uhlir, A. (1956). Electrolytic shaping of germanium and silicon. *Bell Labs Technical Journal*, *35*(2), 333–347.

664. Umezu, I., Yoshida, K., Sugimura, A., Inokuma, T., Hasegawa, S., Wakayama, Y., ... & Yoshida, T. (2000). A comparative study of the photoluminescence properties of a-SiOx: H film and silicon nanocrystallites. *Journal of Non-Crystalline Solids*, *266*, 1029–1032.

665. Vallaitis, T., Koos, C., Bonk, R., Freude, W., Laemmlin, M., Meuer, C., & Leuthold, J. (2008). Slow and fast dynamics of gain and phase in a quantum dot semiconductor optical amplifier. *Optics Express*, *16*(1), 170–178.

666. Van Sark, W. G. J. H. M., Meijerink, A., Schropp, R. E. I., Van Roosmalen, J. A. M., & Lysen, E. H. (2004). Modeling improvement of spectral response of solar cells by deployment of spectral converters containing semiconductor nanocrystals. *Semiconductors*, *38*(8), 962–969.

667. Vavilov, V. S. (1959). On photo-ionization by fast electrons in germanium and silicon. *Journal of Physics and Chemistry of Solids*, *8*, 223–226.

668. Vinciguerra, V., Franzo, G., Priolo, F., Iacona, F., & Spinella, C. (2000). Quantum confinement and recombination dynamics in silicon nanocrystals embedded in Si/SiO 2 superlattices. *Journal of Applied Physics*, *87*(11), 8165–8173.

669. Vinodh, M. S., Mohan, S., Subaskar, P., Bhojraj, H., & Narayanamurthy, H. (2003, October). Indium tin oxide coatings on flexible and rigid substrates for application in smart windows. In *Smart Materials, Structures, and Systems* (Vol. 5062, pp. 780–788). International Society for Optics and Photonics.

670. Vogel, R., Pohl, K., & Weller, H. (1990). Sensitization of highly porous, polycrystalline TiO2 electrodes by quantum sized CdS. *Chemical Physics Letters*, *174*(3–4), 241–246.

671. Voiry, D., Yamaguchi, H., Li, J., Silva, R., Alves, D. C., Fujita, T.,

& Chhowalla, M. (2013). Enhanced catalytic activity in strained chemically exfoliated WS 2 nanosheets for hydrogen evolution. *Nature Materials*, *12*(9), 850.

672. Vossmeyer, T., Katsikas, L., Giersig, M., Popovic, I. G., Diesner, K., Chemseddine, A., & Weller, H. (1994). CdS nanoclusters: synthesis, characterization, size dependent oscillator strength, temperature shift of the excitonic transition energy, and reversible absorbance shift. *The Journal of Physical Chemistry*, *98*(31), 7665–7673.
673. Wadia, C., Alivisatos, A. P., & Kammen, D. M. (2009). Materials availability expands the opportunity for large-scale photovoltaics deployment. *Environmental Science & Technology*, *43*(6), 2072–2077.
674. Wan, Z., Patterson, R., Huang, S., Green, M., & Conibeer, G. (2011). Ultra-thin silicon nitride barrier implementation for Si nano-crystals embedded in amorphous silicon carbide matrix with hybrid superlattice structure. *EPL (Europhysics Letters)*, *95*(6), 67006.
675. Wang, F., Pang, S., Wang, L., Li, Q., Kreiter, M., & Liu, C. Y. (2010). One-step synthesis of highly luminescent carbon dots in noncoordinating solvents. *Chemistry of Materials*, *22*(16), 4528–4530.
676. Wang, H., Gao, P., Wang, Y., Guo, J., Zhang, K. Q., Du, D., & Zou, G. (2015). Fluorescently tuned nitrogen-doped carbon dots from carbon source with different content of carboxyl groups. *APL Materials*, *3*(8), 086102.
677. Wang, H., Kubo, T., Nakazaki, J., Kinoshita, T., & Segawa, H. (2013). PbS-quantum-dot-based heterojunction solar cells utilizing ZnO nanowires for high external quantum efficiency in the near-infrared region. *The Journal of Physical Chemistry Letters*, *4*(15), 2455–2460.
678. Wang, H., Pei, Y., LaLonde, A. D., & Snyder, G. J. (2011). Heavily Doped p-Type PbSe with High Thermoelectric Performance: An Alternative for PbTe. *Advanced Materials*, *23*(11), 1366–1370.
679. Wang, H., Wang, Y., Guo, J., Su, Y., Sun, C., Zhao, J., & Zou, G. (2015). A new chemosensor for Ga 3+ detection by fluorescent nitrogen-doped graphitic carbon dots. *Rsc Advances*, *5*(17), 13036–13041.
680. Wang, J. T. W., Ball, J. M., Barea, E. M., Abate, A., Alexander-Webber, J. A., Huang, J., & Nicholas, R. J. (2013). Low-temperature processed electron collection layers of graphene/TiO_2 nanocomposites in thin film perovskite solar cells. *Nano letters*, *14*(2), 724–730.
681. Wang, J., Mora-Seró, I., Pan, Z., Zhao, K., Zhang, H., Feng, Y., &

Bisquert, J. (2013). Core/shell colloidal quantum dot exciplex states for the development of highly efficient quantum-dot-sensitized solar cells. *Journal of the American Chemical Society*, *135*(42), 15913–15922.

682. Wang, K., Zhao, W., Liu, J., Niu, J., Liu, Y., Ren, X., & Liu, S. F. (2017). CO2 Plasma-Treated TiO2 Film as an Effective Electron Transport Layer for High-Performance Planar Perovskite Solar Cells. *ACS Applied Materials & Interfaces*, *9*(39), 33989–33996.
683. Wang, L., Wang, Y., Xu, T., Liao, H., Yao, C., Liu, Y., & Wu, M. (2014). Gram-scale synthesis of single-crystalline graphene quantum dots with superior optical properties. *Nature communications*, *5*, 5357.
684. Wang, P., Jiang, T., Zhu, C., Zhai, Y., Wang, D., & Dong, S. (2010). One-step, solvothermal synthesis of graphene-CdS and graphene-ZnS quantum dot nanocomposites and their interesting photovoltaic properties. *Nano Research*, *3*(11), 794–799.
685. Wang, X., Koleilat, G. I., Tang, J., Liu, H., Kramer, I. J., Debnath, R., & Sargent, E. H. (2011). Tandem colloidal quantum dot solar cells employing a graded recombination layer. *Nature Photonics*, *5*(8), 480.
686. Wang, X., Ren, X., Kahen, K., Hahn, M. A., Rajeswaran, M., Maccagnano-Zacher, S., & Krauss, T. D. (2009). Non-blinking semiconductor nanocrystals. *Nature*, *459*(7247), 686.
687. Wang, Y., Yao, H. B., Wang, X. H., & Yu, S. H. (2011). One-pot facile decoration of CdSe quantum dots on graphene nanosheets: novel graphene-CdSe nanocomposites with tunable fluorescent properties. *Journal of Materials Chemistry*, *21*(2), 562–566.
688. Wei, D., Andrew, P., & Ryhänen, T. (2010). Electrochemical photovoltaic cells—review of recent developments. *Journal of Chemical Technology and Biotechnology*, *85*(12), 1547–1552.
689. Weidman, M. C., Yager, K. G., & Tisdale, W. A. (2015). Correction to Interparticle Spacing and Structural Ordering in Superlattice PbS Nanocrystal Solids Undergoing Ligand Exchange. *Chemistry of Materials*, *27*(21), 7546–7546.
690. Westbergh, P., Gustavsson, J. S., Kogel, B., Haglund, A., Larsson, A., Mutig, A., & Joel, A. (2010). 40 Gbit/s error-free operation of oxide-confined 850 nm VCSEL. *Electronics Letters*, *46*(14), 1014–1016.
691. Westland, D. J., Ryan, J. F., Scott, M. D., Davies, J. I., & Riffat, J. R. (1988). Hot carrier energy loss rates in GaInAs/InP quantum wells. In *Hot Carriers in Semiconductors* (pp. 431–434).

692. Wheeler, L. M., Nichols, A. W., Chernomordik, B. D., Anderson, N. C., Beard, M. C., & Neale, N. R. (2016). All-inorganic Germanium nanocrystal films by cationic ligand exchange. *Nano Letters*, *16*(3), 1949–1954.

693. Williams, K. A., Thompson, M. G., & White, I. H. (2004). Long-wavelength monolithic mode-locked diode lasers. *New Journal of Physics*, *6*(1), 179.

694. Wilson, M. W., Rao, A., Ehrler, B., & Friend, R. H. (2013). Singlet exciton fission in polycrystalline pentacene: from photophysics toward devices. *Accounts of Chemical Research*, *46*(6), 1330–1338.

695. Wilson, W. L., Szajowski, P. F., & Brus, L. E. (1993). Quantum confinement in size-selected, surface-oxidized silicon nanocrystals. *Science*, *262*(5137), 1242–1244.

696. Wolf, M., Brendel, R., Werner, J. H., & Queisser, H. J. (1998). Solar cell efficiency and carrier multiplication in Si 1− x Ge x alloys. *Journal of Applied Physics*, *83*(8), 4213–4221.

697. Wolfsegger, C., Stierstorfer, J., & Aubrey, C. (2007). Solar Generation IV 2007.Solar electricity for over 1 billion people and 2 million jobs by 2020.

698. Woo, J. Y., Ko, J. H., Song, J. H., Kim, K., Choi, H., Kim, Y. H., & Jeong, S. (2014). Ultrastable PbSe nanocrystal quantum dots via in situ formation of atomically thin halide adlayers on PbSe (100). *Journal of the American Chemical Society*, *136*(25), 8883–8886.

699. Wood, V., & Bulović, V. (2010). Colloidal quantum dot light-emitting devices. *Nano Reviews*, *1*(1), 5202.

700. Wood, V., Halpert, J. E., Panzer, M. J., Bawendi, M. G., & Bulovic, V. (2009). Alternating current driven electroluminescence from ZnSe/ZnS: Mn/ZnS nanocrystals. *Nano Letters*, *9*(6), 2367–2371.

701. Wood, V., Panzer, M. J., Bozyigit, D., Shirasaki, Y., Rousseau, I., Geyer, S., & Bulovic, V. (2011). Electroluminescence from nanoscale materials via field-driven ionization. *Nano Letters*, *11*(7), 2927–2932.

702. Wood, V., Panzer, M. J., Caruge, J. M., Halpert, J. E., Bawendi, M. G., & Bulovic, V. (2009). Air-stable operation of transparent, colloidal quantum dot-based LEDs with a unipolar device architecture. *Nano Letters*, *10*(1), 24–29.

703. Wood, V., Panzer, M. J., Halpert, J. E., Caruge, J. M., Bawendi, M. G., & Bulovic, V. ACS Nano 3, 3581 (2009). *CrossRef| PubMed| CAS|*

Web of Science® Times Cited, 75.

704. Wu, K., Hill, L. J., Chen, J., McBride, J. R., Pavlopolous, N. G., Richey, N. E., & Lian, T. (2015). Universal length dependence of rod-to-seed exciton localization efficiency in type I and quasi-type II CdSe@ CdS nanorods. *ACS Nano*, *9*(4), 4591–4599.

705. Wu, M., Congreve, D. N., Wilson, M. W., Jean, J., Geva, N., Welborn, M., & Baldo, M. A. (2016). Solid-state infrared-to-visible upconversion sensitized by colloidal nanocrystals. *Nature Photonics*, *10*(1), 31.

706. Wu, Z. L., Zhang, P., Gao, M. X., Liu, C. F., Wang, W., Leng, F., & Huang, C. Z. (2013). One-pot hydrothermal synthesis of highly luminescent nitrogen-doped amphoteric carbon dots for bioimaging from Bombyx mori silk–natural proteins. *Journal of Materials Chemistry B*, *1*(22), 2868–2873.

707. Würfel, P. (1997). Solar energy conversion with hot electrons from impact ionisation. *Solar Energy Materials and Solar Cells*, *46*(1), 43–52.

708. Würfel, P., Brown, A. S., Humphrey, T. E., & Green, M. A. (2005). Particle conservation in the hot-carrier solar cell. *Progress in Photovoltaics: Research and Applications*, *13*(4), 277–285.

709. Xie, Z., Wang, F., & Liu, C. Y. (2012). Organic–inorganic hybrid functional carbon dot gel glasses. *Advanced materials*, *24*(13), 1716–1721.

710. Xu, L., Yao, Y., Bronstein, N. D., Li, L., Alivisatos, A. P., & Nuzzo, R. G. (2016). Enhanced photon collection in luminescent solar concentrators with distributed Bragg reflectors. *ACS Photonics*, *3*(2), 278–285.

711. Xu, W., Tan, F., Liu, Q., Liu, X., Jiang, Q., Wei, L., & Wang, Z. (2017). Efficient PbS QD solar cell with an inverted structure. *Solar Energy Materials and Solar Cells*, *159*, 503–509.

712. Xu, X., Ray, R., Gu, Y., Ploehn, H. J., Gearheart, L., Raker, K., & Scrivens, W. A. (2004). Electrophoretic analysis and purification of fluorescent single-walled carbon nanotube fragments. *Journal of the American Chemical Society*, *126*(40), 12736–12737.

713. Yan, X., Cui, X., & Li, L. S. (2010). Synthesis of large, stable colloidal graphene quantum dots with tunable size. *Journal of the American Chemical Society*, *132*(17), 5944–5945.

714. Yan, X., Cui, X., Li, B., & Li, L. S. (2010). Large, solution-processable graphene quantum dots as light absorbers for photovoltaics. *Nano*

letters, *10*(5), 1869–1873.

715. Yan, X., Li, X., Yan, Z. and Komarneni, S. (2014). Porous carbons prepared by direct carbonization of MOFs for supercapacitors. *Applied Surface Science, 308*, 306–310.

716. Yang, C. S., Lin, C. J., Kuei, P. Y., Horng, S. F., Hsu, C. C. H., & Liaw, M. C. (1997). Quantum size effects on photoluminescence from Si nanocrystals in PECVD silicon-rich-oxide. *Applied Surface Science, 113*, 116–120.

717. Yang, L., Tabachnyk, M., Bayliss, S. L., Böhm, M. L., Broch, K., Greenham, N. C., & Ehrler, B. (2014). Solution-processable singlet fission photovoltaic devices. *Nano Letters*, *15*(1), 354–358.

718. Yang, S. T., Cao, L., Luo, P. G., Lu, F., Wang, X., Wang, H., & Sun, Y. P. (2009). Carbon dots for optical imaging in vivo. *Journal of the American Chemical Society*, *131*(32), 11308–11309.

719. Yang, Y., Cui, J., Zheng, M., Hu, C., Tan, S., Xiao, Y., & Liu, Y. (2012). One-step synthesis of amino-functionalized fluorescent carbon nanoparticles by hydrothermal carbonization of chitosan. *Chemical Communications*, *48*(3), 380–382.

720. Yang, Y., Li, J., Lin, L., & Peng, X. (2015). An efficient and surface-benign purification scheme for colloidal nanocrystals based on quantitative assessment. *Nano Research*, *8*(10), 3353–3364.

721. Yang, Y., Ostrowski, D. P., France, R. M., Zhu, K., Van De Lagemaat, J., Luther, J. M., & Beard, M. C. (2016). Observation of a hot-phonon bottleneck in lead-iodide perovskites. *Nature Photonics*, *10*(1), 53–59.

722. Yang, Y., Qu, L., Dai, L., Kang, T. S., & Durstock, M. (2007). Electrophoresis coating of titanium dioxide on aligned carbon nanotubes for controlled syntheses of photoelectronic nanomaterials. *Advanced Materials*, *19*(9), 1239–1243.

723. Yang, Y., Rodriguez-Cordoba, W., Xiang, X., & Lian, T. (2011). Strong electronic coupling and ultrafast electron transfer between PbS quantum dots and TiO2 nanocrystalline films. *Nano Letters*, *12*(1), 303–309.

724. Ye, M., Gao, X., Hong, X., Liu, Q., He, C., Liu, X., & Lin, C. (2017). Recent advances in quantum dot-sensitized solar cells: insights into photoanodes, sensitizers, electrolytes and counter electrodes. *Sustainable Energy & Fuels*, *1*(6), 1217–1231.

725. Ye, R., Xiang, C., Lin, J., Peng, Z., Huang, K., Yan, Z., & Ceriotti, G.

(2013). Coal as an abundant source of graphene quantum dots. *Nature communications*, *4*, 2943.

726. Yin, Y., & Alivisatos, A. P. (2004). Colloidal nanocrystal synthesis and the organic–inorganic interface. *Nature*, *437*(7059), 664.Yin, Y., Erdonmez, C., Aloni, S., & Alivisatos, A. P. (2006). Faceting of nanocrystals during chemical transformation: From solid silver spheres to hollow gold octahedra. *Journal of the American Chemical Society*, *128*(39), 12671–12673.

727. Yu, D., Wang, C., & Guyot-Sionnest, P. (2003). n-Type conducting CdSe nanocrystal solids. *Science*, *300*(5623), 1277–1280.

728. Yu, H., Zhang, H., Huang, H., Liu, Y., Li, H., Ming, H., & Kang, Z. (2012). ZnO/carbon quantum dots nanocomposites: one-step fabrication and superior photocatalytic ability for toxic gas degradation under visible light at room temperature. *New Journal of Chemistry*, *36*(4), 1031–1035.

729. Yu, W. W., Qu, L., Guo, W., & Peng, X. (2003). Experimental determination of the extinction coefficient of CdTe, CdSe, and CdS nanocrystals. *Chemistry of Materials*, *15*(14), 2854–2860.

730. Yuan, M., Liu, M., & Sargent, E. H. (2016). Colloidal quantum dot solids for solution-processed solar cells. *Nature Energy*, *1*(3), 16016.

731. Yuan, M., Zhong, R., Gao, H., Li, W., Yun, X., Liu, J., & Zhang, F. (2015). One-step, green, and economic synthesis of water-soluble photoluminescent carbon dots by hydrothermal treatment of wheat straw, and their bio-applications in labeling, imaging, and sensing. *Applied Surface Science*, *355*, 1136–1144.

732. Yun, J. S., Yang, K. S., & Kim, D. H. (2011). Multifunctional polydiacetylene-Graphene nanohybrids for biosensor application. *Journal of Nanoscience and Nanotechnology*, *11*(7), 5663–5669.

733. Zarghami, M. H., Liu, Y., Gibbs, M., Gebremichael, E., Webster, C., & Law, M. (2010). p-Type PbSe and PbS quantum dot solids prepared with short-chain acids and diacids. *Acs Nano*, *4*(4), 2475–2485.

734. Zatryb, G., Podhorodecki, A., Hao, X. J., Misiewicz, J., Shen, Y. S., & Green, M. A. (2010). Quantitative evaluation of boron-induced disorder in multilayers containing silicon nanocrystals in an oxide matrix designed for photovoltaic applications. *Optics Express*, *18*(21), 22004–22009.

735. Zatryb, G., Podhorodecki, A., Hao, X. J., Misiewicz, J., Shen, Y.

S., & Green, M. A. (2011). Correlation between stress and carrier nonradiative recombination for silicon nanocrystals in an oxide matrix. *Nanotechnology*, *22*(33), 335703.

736. Zedan, A. F., Sappal, S., Moussa, S., & El-Shall, M. S. (2010). Ligand-controlled microwave synthesis of cubic and hexagonal CdSe nanocrystals supported on graphene. Photoluminescence quenching by graphene. *The Journal of Physical Chemistry C*, *114*(47), 19920–19927.
737. Zemnukhova, L. A., Panasenko, A. E., Fedorishcheva, G. A., Ziatdinov, A. M., Polyakova, N. V., & Kuryavyi, V. G. (2012). Properties of silicon prepared from plant raw materials. *Inorganic Materials*, *48*(10), 971–976.
738. Zhang, G., Finefrock, S., Liang, D., Yadav, G. G., Yang, H., Fang, H., & Wu, Y. (2011). Semiconductor nanostructure-based photovoltaic solar cells. *Nanoscale*, *3*(6), 2430–2443.
739. Zhang, H., Hyun, B. R., Wise, F. W., & Robinson, R. D. (2012). A generic method for rational scalable synthesis of monodisperse metal sulfide nanocrystals. *Nano Letters*, *12*(11), 5856–5860.
740. Zhang, J., Chernomordik, B. D., Crisp, R. W., Kroupa, D. M., Luther, J. M., Miller, E. M., & Beard, M. C. (2015). Preparation of Cd/Pb chalcogenide heterostructured janus particles via controllable cation exchange. *ACS Nano*, *9*(7), 7151–7163.
741. Zhang, J., Gao, J., Church, C. P., Miller, E. M., Luther, J. M., Klimov, V. I., & Beard, M. C. (2014). PbSe quantum dot solar cells with more than 6% efficiency fabricated in ambient atmosphere. *Nano Letters*, *14*(10), 6010–6015.
742. Zhang, J., Gao, J., Miller, E. M., Luther, J. M., & Beard, M. C. (2013). Diffusion-controlled synthesis of PbS and PbSe quantum dots with in situ halide passivation for quantum dot solar cells. *ACS Nano*, *8*(1), 614–622.
743. Zhang, P., Li, W., Zhai, X., Liu, C., Dai, L., & Liu, W. (2012). A facile and versatile approach to biocompatible "fluorescent polymers" from polymerizable carbon nanodots. *Chemical Communications*, *48*(84), 10431–10433.
744. Zhang, Q., Wang, H., Liu, W., Wang, H., Yu, B., Zhang, Q., & Chen, G. (2012). Enhancement of thermoelectric figure-of-merit by resonant states of aluminium doping in lead selenide. *Energy & Environmental*

Science, *5*(1), 5246–5251.

745. Zhang, X., Hui, J., Yang, B., Yang, Y., Fan, D., Liu, M., & Wei, Y. (2013). PEGylation of fluoridated hydroxyapatite (FAp): Ln 3+ nanorods for cell imaging. *Polymer Chemistry*, *4*(15), 4120–4125.
746. Zhang, X., Wang, S., Zhu, C., Liu, M., Ji, Y., Feng, L., & Wei, Y. (2013). Carbon-dots derived from nanodiamond: Photoluminescence tunable nanoparticles for cell imaging. *Journal of colloid and interface science*, *397*, 39–44.
747. Zhang, X., Zhang, Y., Wang, Y., Kalytchuk, S., Kershaw, S. V., Wang, Y., & Cui, T. (2013). Color-switchable electroluminescence of carbon dot light-emitting diodes. *ACS nano*, *7*(12), 11234–11241.
748. Zhang, Y. Q., Ma, D. K., Zhang, Y. G., Chen, W., & Huang, S. M. (2013). N-doped carbon quantum dots for TiO2-based photocatalysts and dye-sensitized solar cells. *Nano Energy*, *2*(5), 545–552.
749. Zhang, Z., Hao, J., Zhang, J., Zhang, B. and Tang, J. (2012). Protein as the source for synthesizing fluorescent carbon dots by a one-pot hydrothermal route. *RSC Advances, 2*(23), 8599–8601.
750. Zhao, A., Zhao, C., Li, M., Ren, J., & Qu, X. (2014). Ionic liquids as precursors for highly luminescent, surface-different nitrogen-doped carbon dots used for label-free detection of Cu2+/Fe3+ and cell imaging. *Analytica Chimica Acta*, *809*, 128–133.
751. Zhao, K., Pan, Z., Mora-Seró, I., Cánovas, E., Wang, H., Song, Y., & Zhong, X. (2015). Boosting power conversion efficiencies of quantum-dot-sensitized solar cells beyond 8% by recombination control. *Journal of the American Chemical Society*, *137*(16), 5602–5609.
752. Zhao, Q. L., Zhang, Z. L., Huang, B. H., Peng, J., Zhang, M., & Pang, D. W. (2008). Facile preparation of low cytotoxicity fluorescent carbon nanocrystals by electrooxidation of graphite. *Chemical Communications*, (41), 5116–5118.
753. Zhao, Y., & Burda, C. (2012). Development of plasmonic semiconductor nanomaterials with copper chalcogenides for a future with sustainable energy materials. *Energy & Environmental Science*, *5*(2), 5564–5576.
754. Zheng, M., Liu, S., Li, J., Qu, D., Zhao, H., Guan, X., & Sun, Z. (2014). Integrating oxaliplatin with highly luminescent carbon dots: an unprecedented theranostic agent for personalized medicine. *Advanced Materials*, *26*(21), 3554–3560.

755. Zhitomirsky, D., Voznyy, O., Levina, L., Hoogland, S., Kemp, K. W., Ip, A. H., & Sargent, E. H. (2014). Engineering colloidal quantum dot solids within and beyond the mobility-invariant regime. *Nature Communications*, *5*, 3803.

756. Zhou, J., Booker, C., Li, R., Zhou, X., Sham, T. K., Sun, X., & Ding, Z. (2007). An electrochemical avenue to blue luminescent nanocrystals from multiwalled carbon nanotubes (MWCNTs). *Journal of the American Chemical Society*, *129*(4), 744–745.

757. Zhou, J., Sheng, Z., Han, H., Zou, M. and Li, C. (2012). Facile synthesis of fluorescent carbon dots using watermelon peel as a carbon source. *Materials Letters, 66*(1), 222–224.

758. Zhou, X., Huang, X., Qi, X., Wu, S., Xue, C., Boey, F. Y., & Zhang, H. (2009). In situ synthesis of metal nanoparticles on single-layer graphene oxide and reduced graphene oxide surfaces. *The Journal of Physical Chemistry C*, *113*(25), 10842–10846.

759. Zhou, X., Zhang, Y., Wang, C., Wu, X., Yang, Y., Zheng, B., Wu, H., Guo, S. and Zhang, J. (2012). Photo-Fenton Reaction of Graphene Oxide: A New Strategy to Prepare Graphene Quantum Dots for DNA Cleavage. *ACS Nano, 6*(8), 6592–6599.

760. Zhu, H., Song, N., & Lian, T. (2013). Charging of quantum dots by sulfide redox electrolytes reduces electron injection efficiency in quantum dot sensitized solar cells. *Journal of the American Chemical Society*, *135*(31), 11461–11464.

761. Zhu, S., Meng, Q., Wang, L., Zhang, J., Song, Y., Jin, H., & Yang, B. (2013). Highly photoluminescent carbon dots for multicolor patterning, sensors, and bioimaging. *Angewandte Chemie*, *125*(14), 4045–4049.

762. Zhu, Z., Ma, J., Wang, Z., Mu, C., Fan, Z., Du, L., & Yang, S. (2014). Efficiency enhancement of perovskite solar cells through fast electron extraction: the role of graphene quantum dots. *Journal of the American Chemical Society*, *136*(10), 3760–3763.

INDEX

M

N

O

P

Q

R

S

T

V

X

Printed and bound by PG in the USA